THE WORLD IS NOT FOR SALE

THE WORLD IS NOT FOR SALE

Farmers Against Junk Food

◆

JOSÉ BOVÉ
and
FRANÇOIS DUFOUR

Interviewed by Gilles Luneau

Translated by Anna de Casparis

VERSO
London • New York

First published by Verso 2001
Paperback edition first published by Verso 2002
© Verso 2001, 2002
Translation © Anna de Casparis 2001, 2002
First published as *Le monde n'est pas une marchandise*
© Editions La Découverte & Syros, Paris, 2000

1 3 5 7 9 10 8 6 4 2

Verso
UK: 6 Meard Street, London W1F 0EG
US: 180 Varick Street, New York, NY 10014–4606
www.versobooks.com

Verso is the imprint of New Left Books

ISBN 1–85984–405–7

British Library Cataloguing in Publication Data
A catalogue record for this book is available from the British Library

Library of Congress Cataloging-in-Publication Data
A catalog record for this book is available from the Library of Congress

Typeset in 11/14pt Perpetua by SetSystems Ltd, Saffron Walden, Essex
Printed by Biddles Ltd, Guildford and King's Lynn

CONTENTS

♥

CONTENTS

FOREWORD

When José Bové first came to world attention after the symbolic dismantling of a McDonald's in Millau, the action was widely misrepresented as anti-American and protectionist, or sometimes simply as French food snobbery. Less than two years later, José Bové and François Dufour are at the forefront of an international pro-democracy movement. With Mad Cow and foot-and-mouth disease ravaging the world agricultural markets, seemingly abstract debates about the centralization, medicalization and genetic manipulation of food production have become both pressing and personal. Every time we sit down to eat, or go to the supermarket, it becomes more difficult to avert our eyes from the true cost of treating food like an industrial commodity. Today, millions are now asking the same question José Bové demanded that day in Millau: who gets to make decisions about the quality and integrity of the food we eat – citizens, or faceless international trade institutions?

As doubts deepen about the viability of industrial agriculture and fast food, José Bové and François Dufour have come to stand for a way of life in which our relationship to food and nature is grounded in respect. For them, food is more than bodily fuel; it is ritual, relationship, family, love, tradition, and so much else. These values cannot be measured in economic terms: how do you measure the value of a family meal? How do you calculate the social contribution of unproductive time?

Even though the market cannot accommodate these ideas, most of us understand them intuitively, which is why Bové and Dufour have become folk heroes, fighting in defense of all that cannot be counted, but that counts more than anything else.

Theirs is not a nationalist battle, nor is it only about food. Bové's and Dufour's message has resonance everywhere battles are fought for the right to local democracy and cultural diversity in a world governed increasingly by the principles that govern McDonald's: the same fare everywhere you go. It is about the right to distinct, uncommodified spaces – cultural activities, rituals, pieces of our ecology, ideas, life itself – that are not for sale.

Naomi Klein,
author of No Logo

PREFACE

Our story begins in Millau, southwest France, on 12 August 1999: the day local sheep farmers dismantled a partially constructed McDonald's in their town. They did this in response to America's attack on their main export: Roquefort cheese, one of the highest-quality French products, had become a victim of American trade barriers introduced in retaliation for Europe's refusal to import hormone-fed beef. The symbolism of the action was powerful: wholesome food against tasteless muck; agricultural workers versus the power of the multinationals. The demonstration went beyond mere trade-union issues: consumers joined the farmers in their protest, and local townspeople made up half the demonstration. French people respond when there is an attack on good food or its guardians, the farmers.

In sentencing José Bové and his fellow union members to imprisonment, the local judge ironically gave the protesters a big boost, attracting major media attention to their cause. Rather than responding mildly to a good-natured disruption, it appears that she followed orders from higher up. She decided, along with the local prefecture, that the damage to the McDonald's was twenty times greater than it actually was. And she showed complete disregard for what had been a genuine expression of popular opinion against heavy-handed American action.

Five farmers were sent to prison. As many townspeople as farmers attended the demonstration, and as many women as men, but five male farmers were singled out and locked up. Not all the sheep farmers on the demonstration were union members, but all five men arrested belonged to the radical Farmers' Confederation. Many suspected a stitch-up.

Perhaps it is not surprising that the public reaction to a prison sentence handed out for the offence of defending the quality of food was so strong. Consumers, after all, have been through a number of food scares – mad cow disease, nitrates in the water supply, genetically modified food, listeria, heavy-metal residues – and are already sensitive to the ever-increasing dangers of intensive agriculture.

In these media-orientated days, the struggle against the food scandals needs a charismatic figurehead, and the Millau protesters found theirs: José Bové, a blue-eyed, blond-moustachioed, local farmer who has subsequently become world famous as an anti-junk-food hero. From his prison cell, he witnessed demands for his release burgeoning across France, before spreading through Europe and, eventually, across the Atlantic to North America. This growth of the campaign owed a lot to the existence of the World Wide Web. Though they were not conscious of it, the sheep farmers had acted as a catalyst for a response to a global problem.

When you realize you no longer know what children are being fed in school cafeterias, when the meat they eat may contain infected brain matter, when local breeds and plant life are vanishing, when any certainty about food has melted away, then you support people like José Bové.

When Africans are further impoverished by European Union agricultural exports or American food subsidies, then you support campaigns like those of the farmers in Millau. If you are a disaffected American who sees the action of Bové and his comrades as David taking on Goliath, you take the side of David.

On the steps of the court Bové was photographed, handcuffed arms held above his head, smiling faintly. This image has become a historic symbol; it illustrates a world where we live in chains, where revolt is both necessary and legitimate. José Bové refused to pay bail to escape going to jail, and his fellow prisoners gave him an ovation. In the end, it was supporters from around the world who sent cheques by the thousand to set him free. Once he was out of prison, the struggle had moved beyond a local dispute in southwest France. It was no longer just a fight against unfair Customs duties, or even against bad food. It had become a massive campaign against the murky dealings of international trade.

Preparations for the demonstration at Seattle in November 2000 were under way. The World Trade Organization meeting was accompanied by a counter-summit, drawing tens of thousands of opponents to a money-dominated world. François Dufour, General National Secretary of the French Farmers' Confederation, and his fellow trade unionists, including José Bové, went to America a little ahead of the scheduled events in Seattle. They spent ten days travelling across the USA, east to west, meeting farmers, consumers and environmentalists. Their message was clear: they were not anti-American, but they demanded fair trade in the world market. As they toured across the country, the meetings got bigger and bigger; their reputation ran ahead of them. As a result of their visit, America heard the voice of its own small farmers, who took advantage of the media coverage attracted by the visitors to put out an appeal against genetically modified foods and similar agricultural malpractices in the USA. People were urged to join the protests in Seattle.

On Tuesday 30 November, the centre of Seattle was paralysed by tens of thousands of demonstrators, including farmers from eighty different countries. They prevented the WTO meeting from opening. After a curfew had been imposed, the Conference eventually went ahead under the protection of the national guard, but it failed to conclude its business. An agreement would surely have been reached had it not been for the actions of the international demonstrators. The new idea – writ large on the streets, and spread through the Internet – was that, as capitalism at its most rampant and ruthless was sweeping the world, a global stance in favour of democracy was now required.

The demonstrations on the streets were only a beginning. The heroes of this book, Bové and Dufour, discuss what is required to exercise popular control over bodies like the WTO. They ask why whole areas of the economy have been globalized without significant public discussion or approval. They examine, in particular, how these issues have a bearing on the field of agriculture in which they work. In the race for profit, they point out, agribusiness has taken over farming. From its original function of providing food, agriculture has become just another way to make more money, with little or no regard for those who work the land or who eat what is grown.

But, Bové and Dufour argue, food is not a mere commodity: eating is an intimate, daily activity, a source of pleasure, a means of survival, and a critical aspect of the way in which we relate to the earth. Food has its rituals in every culture, creed, religion and philosophy. Wheat, maize and rice are more than just crops. They are the outcome of a fusion of sun, water and soil. In eating, humans inscribe themselves in the cycles of the universe, and this is far more profound and basic than just making money. Wheat was growing long before coins were cast. If we want to have a say in what we eat, we must control global trade. This was the lesson of José Bové and François Dufour in autumn 1999. Their message has been taken up from Millau to Seattle: sustainable or traditional agriculture and wholesome food are worth fighting for, because the world is not for sale – and neither are we.

Gilles Luneau

LIST OF ABBREVIATIONS

AMI   Accord multilatéral sur l'investissement/Multilateral Agreement on Investment

ATTAC   Association pour la taxation des transactions financières pour l'aide aux citoyens: a pro-citizens taxation pressure group

CCJA   Centre cantonal des jeunes agriculteurs/Area Centre for Young Farmers

CDJA   Centre départemental des jeunes agriculteurs/Departmental Centre for Young Farmers

CDOA   Commission consultative départementale d'orientation de l'agriculture/Departmental Consultative Commission for Agriculture

CEDAPA Centre d'études pour un développement agricole plus autonome/Study Centre for the Development of More Autonomous Agriculture

CFDT   Confédération française démocratique du travail: a major French trade union

CGT   Conféderation générale du travail: a major French trade union

CIVAM   Centre d'initiatives pour valoriser l'agriculture et le milieu rural/ Organization for the Promotion of Farming and Rural Life

CJA   local level of the CDJA

CNDFS   Coordination nationale de défense des semences fermières/National Co-ordination for the Protection of Farmers' Seeds, an umbrella organization which brings together several farmers' and agricultural workers' unions

CNJA   Centre national des jeunes agriculteurs/National Centre for Young Farmers

CNSTP   Confédération nationale des syndicats de travailleurs paysans/ National Confederation of Farm Workers' Unions

CNUCED   Conférence des Nations unies pour le commerce et le développement/United Nations Conference on Trade and Development

CPE   Coordination paysanne européenne/European Farmers' Co-ordination

CPNT   Chasse, pêche, nature et tradition/Hunting, Fishing, Nature and Tradition

CRS   Compagnies républicaines de sécurité: French state riot police

CUMA   Coopérative d'utilisation de matériel agricole/Co-operative for the Use of Agricultural Equipment

FNSEA   Fédération nationale des syndicats d'exploitants agricoles/Union of Farm Labourers' National Confederation: the main national farmers' union, organized on a regional (FRSEA) and Departmental (FDSEA) basis

FNSP   Fédération nationale des syndicats de paysans/National Federation of Farmers' Unions

GAEC   Groupement agricole d'exploitation en commun: an agricultural co-operative

GATT   General Agreement on Tariffs and Trade

IMF   International Monetary Fund

JAC   Jeunesse agricole catholique/Young Catholic Farmers

ORD   Organe de règlement des différends/Organization for Regulating Differentials

LIST OF ABBREVIATIONS

RPR	Rassemblement pour la république: a right-wing Gaullist political organization
SCTL	Société civile des terres du Larzac/Larzac Land Associates
SPLB	Syndicat des producteurs de lait de brebis/Union of Ewe's Milk Producers
UDF	Union pour la démocratie française: a centre–right political coalition
WTO	World Trade Organization

PART ONE

TWO FIGHTING FARMERS

1

McDONALD'S: BOVÉ'S STORY

Agence France Presse, 12 August 1999, 11.14 GMT –
McDonald's construction site ransacked by farmers.

Millau, 12 August – A group of farmers ransacked a McDonald's under
construction in Millau (Aveyron), during a demonstration on Thursday
against the American-imposed sanctions against the ban on the import
of hormone-treated beef, according to police sources.

At 11 a.m. on Thursday 12 August 1999, the Union of Ewe's Milk
Producers (SPLB) and the Farmers' Confederation called a rally in front of a
McDonald's which was being built on the site of a former petrol station, on
the road south out of Millau. Three hundred people turned up – half from
the countryside, half from the town.

Was any particular organization from Millau involved in calling the demonstration?

José Bové: No. But whenever we mobilize for a demonstration, we
advertise widely in the press; everyone who wanted to support us was
welcome. In this part of the world, the South Aveyron region of France,
solidarity on the issue of ewe's milk is taken for granted, and is just as likely

to come from the trade unions as the farmers' associations, because Roquefort is crucial to the local economy: 1,300 people are employed in its production, an enormous number for this area. So it was not surprising that so many came from the town. These supporters have been around some twenty years, activists in the Larzac movement (see p. 35). They are not strangers to each other, but veterans of the same struggles. We're talking about a relatively large core which can be mobilized quite quickly: three hundred people in the middle of August. People turned up with their families, so there was a festive atmosphere, with kids having a good time mucking about on the building site.

McDONALD'S: A SYMBOL

How long had you been thinking of taking on McDonald's?

JB: You've got to link McDonald's to the issue of hormone-treated meat. At our Congress in Vesoul in April 1999, we'd already raised the question of preparing ourselves for American retaliation against Europe's ban on the import of hormone-treated beef. In February 1998, the World Trade Organization had condemned the European Union's ban, and given it fifteen months to get its house back in order – that is, to reopen its frontiers. This deadline had expired on 13 May 1999, so the American move came as no surprise. We had already envisaged linking the issue of hormones and McDonald's. What we had not foreseen, however, was that Roquefort, the main produce of the farmers in my area, would be included in the hundred or so products affected by a 100 per cent Customs surcharge on entering the States.

In Washington, the price of Roquefort shot up from $30 a kilo to $60, effectively prohibiting its sale. Around the same time, we found out that a McDonald's was being constructed in Millau. The professional association of Roquefort producers decided to lobby the Minister for Agriculture on 5

August. The Minister, Jean Glavany, informed us that there was nothing he could do, that he was unable to obtain direct financial compensation, that Europe was powerless and there was no other way out. He could promise only to finance publicity for a campaign on the produce affected by the surcharge.

What were the financial losses incurred by the producers of ewe's milk following the American ruling?

JB: We sell 440 tonnes of cheese annually to the States, worth 30 million francs. Given that the cost of the milk is half the value of the Roquefort, the producers are losing 15 million francs; this represents 3 million litres of milk out of the 80 million used in the annual production of Roquefort. As a result, the professional association resorted to new measures: 10,000 pamphlets were printed and distributed across the region, including camp-sites and village fêtes. Posters were put up everywhere, and in Millau itself large banners proclaimed: 'No to the US embargo on Roquefort'. It was in the spirit of these activities that we decided, at a meeting of the SPLB, to pay a visit to McDonald's.

We had harboured the idea for a while. We had first mooted it publicly to the TV journalists on the steps outside the Ministry of Agriculture after our meeting with the Minister: 'If no changes are forthcoming in the next few days, we will have no option but to take on McDonald's, the symbol of industrial food and agriculture.' So we'd already signalled our plans. . . .

How did you prepare for this action, and how did events unfold?

JB: The objective was to have a non-violent but symbolically forceful action, in broad daylight and with the largest possible participation. We wanted the authorities to be fully aware of what was going to happen, so we explained to the police in advance that the purpose of the rally was to dismantle the McDonald's. The police notified the regional government, and they called back to say they would ask the McDonald's manager to provide a billboard or something similar for us to demolish. This, he said, would allow for a symbolic protest. We replied that this idea was ludicrous, and that it remained our intention to dismantle the doors and windows of the building. The police deemed it unnecessary to mount a large presence. We asked them to make sure that the site would be clear of workers, and that no tools were left lying around.

It all happened as we'd envisaged. The only odd thing was the presence of some ten plainclothes police officers armed with cameras. The demo took place and people, including kids, began to dismantle the inside of the building, taking down partitions, some doors, fuseboxes, and some tiles from the roof – they were just nailed down, and came off very easily; in fact the whole building appeared to have been assembled from a kit. The structure was very flimsy. While some people started to repaint the roof of the restaurant, others began loading bits of the structure on to tractor trailers. One of the trailers was a grain carrier. As soon as the trailers were loaded, everyone left the site. Children clambered on the grain wagon and used wooden sticks to bang on the sides, and the whole lot proceeded in the direction of the prefecture, seat of the regional government.

As the procession wound its way through the town, the festive atmosphere was further heightened by the cheering of local people who had gathered to watch us go by. We unloaded everything in front of the prefecture. It was a beautiful day, everyone was having a good time, and many people ended up on the terraces of Millau's restaurants.

When I got home that evening, I was quite unprepared for the way in which the media began to report the protests. The newsflash from Agence France Presse (AFP), for instance, said that the McDonald's was 'ransacked', an approach repeated widely elsewhere. The next morning, France 3, the TV news station in the South Pyrenees region, called asking me to appear live on their programme.

At 7 p.m. I presented myself at the news station. Before my spot there was a short report, some two-and-a-half minutes long, showing the manager of McDonald's on the steps of the prefecture, declaring that a million francs' worth of damage had been caused! The Prefect then proceeded to give an entirely fictitious report of what had happened, claiming that we had hidden behind the children to avoid police intervention. I was completely taken aback, and pointed out that we had acted in broad daylight, and that the Prefect had evidently done some of her in-service training in Corsica, where local government is very corrupt. It was now Friday evening, 13 August. The next day I was due to go on holiday.

On the following Tuesday morning, I woke up to the announcement that some of the participants in the demonstration had been arrested. Apparently forty detectives from Languedoc-Roussillon had been mobilized. They had raided five farms including mine, and the house of the president of the Federation of Grands Causses, an umbrella organization for local groups in Millau.

Of course they found no one at my place, but four farmers were arrested: Jean-Émile Sanchez, Christian Roqueirol, Raymond Fabrègues and Léon Maillé, as well as Jacques Barthélémy, president of the Federation. These five men were taken to the police station in Millau.

At this point, we witnessed a crazy turn of events. Although there was not even a shadow of a militant presence outside the police station, it was being guarded by two squads of CRS. The judge, Nathalie Marty, decreed that it was too dangerous to transfer the men from the police station to the courts, so a temporary courtroom was set up at the station!

In order to justify such a course of action legally, there should have been proof that access to the courts was impossible. In fact access was easy: the whole area was closed off to traffic, and there was only one road to cross to get to the courthouse. It just didn't make sense. At the end of the day, the five men were remanded and charged with conspiracy to cause criminal damage.

AN OVER-THE-TOP REACTION

How do you explain such an exaggerated reaction?

JB: There were obviously several factors at play here. In the first place, the Prefect was very inexperienced; this was her first posting. I also think she was under pressure, as she had been put in her post by the left to counteract the political power of Jean Puech, the UDF President of the General Council. The need to establish respect for law enforcement was also key. This was explicitly asserted by the chief of police in the Department, who proclaimed: 'We must repress, repress, repress!' He was fed up with demonstrations. And finally, there was the deputy mayor, Jacques Godfrain, who cannot stomach the Farmers' Confederation. A hardline local supporter of the RPR, Godfrain had not yet come to terms with having his desk chucked out on the pavement in 1994 because he voted in favour of GATT. The fact that the Farmers' Confederation is very active on the ground upset the political and economic powers in the area. We're all troublemakers!

You went off on holiday with your family, and discovered you'd become public enemy Number 1, with a warrant out for your arrest. How did you react?

JB: I was shocked that we were being treated as criminals. Naturally, I decided to turn myself in, but not before taking full stock of the situation with the national secretary of the Farmers' Confederation and people from the local Roquefort association – and not before alerting the media, so that I could explain that we were not criminals. I was not on the run!

All this took three-and-a-half days. I had to be very careful, as I was on the wanted list and I became aware that my mobile phone was being tapped. So, avoiding the main roads, I went back to the Aveyron, to a farm with only one access lane. I even hid in a crate in the back of a van until we got to the presbytery of a priest friend, where we waited for François Dufour, who was arriving to support the demonstration called in solidarity with those imprisoned.

In the hills above Millau, we improvised a press conference. Present at this conference were François Dufour, Alain Soulié, leader of the SPLB, and our lawyers, François Roux and Marie-Christine Etelin. At about 1 p.m., I presented myself in the courtroom. Judge Marty subjected me to the same line of questioning as before, then dispatched me to the prison in Villeneuve-lès-Maguelones, a suburb of Montpellier.

There was certainly a discrepancy between what happened in Millau and the way some of the papers reported it. It was as if the whole of France wanted to believe there had been a rampage at the McDonald's. I encountered this reaction myself when I interviewed people in markets in the Aveyron. Their reaction was: 'It's right to defend Roquefort, it's right to have a go at McDonald's, but it's not right to wreck everything.' Yet when the farmers were distributing leaflets and explaining that there had been no vandalism, only an organized dismantling, people seemed to be sorry that the action hadn't been more militant.

JB: If you were to read the reports in *Midi Libre* and *Le Monde* the next day, you wouldn't think they were describing the same demonstration. It's really pretty amazing. If there had been any doubts within the Confederation itself, these were dispelled by the Thursday, when François Dufour arrived. Four farmers, members of the union, were still inside, and the Farmers' Confederation got to work. The only non-farmer arrested, Jacques Barthé-lémy, had been released the very same evening under orders not to leave the Millau area; this restriction was lifted only two months later, when the witness who claimed to have seen him ripping out cables didn't even turn up at the trial.

WINNING OVER THE PUBLIC

Despite the devastating headlines in the press, public opinion very quickly came round to your side. In fact, the demonstration soon took on a symbolic importance. How do you analyse this phenomenon? Were you aware of it in prison?

JB: In Millau, support had been forthcoming right from the start. Those who had taken part in the action couldn't understand the selective arrests. Farmers, workers, trade unionists, organizations – everyone supported the ewe's milk producers. A petition went round, calling for all the signatories to be brought to trial: hundreds of people signed, more than had demonstrated on the 12th.

Some local elected representatives – including the Socialist member of the General Council who had participated in the action – expressed their support, and kept up the link between Millau and the government. From my prison cell I was very surprised to see that our actions continued to be shown on TV, as the first or second news item. The confrontation remained in the limelight. The courtroom appearances became a focus for demonstrations, and enabled the mobilization to stay alive. It was not long before the political parties got involved: first the Greens, making a fuss to the Minister, Jean Glavany. Journalists and intellectuals followed suit, using newspaper columns and editorials to call on the politicians to do something about the action and my internment. And suddenly leading figures in the political parties, one after the other, expressed more or less the same sentiments along the lines of 'He shouldn't be inside. I may not agree with him, but he must be freed, because the issues he raises are pertinent.'

From my cell, I watched the snowball effect spreading over the whole country. Public opinion had rallied to our idea of being careful about food and our denunciation of being dictated to by international trade. The political parties attempted to fall into line behind public opinion. Embracing the popular movement, individuals from groups and unions, from the extreme right to the extreme left, nationalists, anti-Americans, opportunists of all sorts, supported us. Public support was so strong that it engulfed even those who had previously condemned the anti-McDonald's action, or kept their distance, encouraging them to engage with 'junk food' and world trade, and even to demand my freedom.

August the 31st was the day of your court appearance in Montpellier, to request bail. It was the day when you were photographed with your arms up in the air, handcuffed — a picture that appeared everywhere.

JB: The handcuffs symbolized my arrest. I realized the impact that the image of my holding them up could have, so the photograph was not accidental. It was one I wanted, posed for, you could say. It certainly helped to extend the mobilization and underline the fact that a legitimate protest movement could not be stifled. For the same reason, I turned down an offer of bail some days later when the judge granted it. You can't buy trade-union freedom, and I felt that I had already paid bail by spending a fortnight in prison.

Without any prompting, hundreds of cheques started to arrive from far and wide, indicating the breadth of our support. I read in *Le Monde* that cheques to 'free the French farmer' were coming from American farmers and consumers. All this seemed very surprising from inside a prison cell. The director of the Confederation of Roquefort Producers, a manager of the Besnier group,[1] stated that he was willing to pay my bail, as in his opinion I 'was more useful on the outside, where I could negotiate the price of milk and where my actions had helped to defend his business'.

1. The Besnier group, known today as Lactalis, is the leading European industrial milk producer (it trades under the brand names Président, Lanquetot, Lepetit, Bridel, Lactel, Valmont). It processes 25 per cent of cow's milk produced in France, and has moved into the production of quality produce, including Roquefort.

How do you explain such a widespread movement? Is it because your milk, your cheese, has been targeted? Is it a legacy of the Larzac struggle? Or the result of so many food scares, starting with mad cow disease, then genetically modified foods, benzodioxin, and residues of septic tanks in the food chain?

JB: I think all these factors contributed. What had been happening in the food chain in the months prior to this was certainly crucial: first the Belgian chickens poisoned by benzodioxin, then the Coca-Cola containing dubious substances, and the numerous problems linked to pig-rearing, a very sensitive issue for us – not to mention the continuing worries about mad cow disease, which have been around for some years. All these combined to create a highly sensitive public.

So our action struck a chord. People understood the issue of Roquefort and hormones. Despite the allegations which reduced the action to an anti-McDonald's one, the symbolism of the demonstration in Millau had been understood, and set into motion the wider movement which the Farmers' Confederation had initiated. Every day there were actions in our cities, increasing the numbers that came to the meetings, and more and more trade unions and political organizations expressed their support.

The resonance of the word 'hormone' was obviously to your great advantage in the swing of public opinion?

JB: Indeed it was. Many people still recalled the scandal in the 1980s following the news that calves were being fed hormones; the story had been uncovered by Bernard Lambert and the Worker-Farmers.[2] The word 'hormone' worries people – so do the initials 'GMO', because they raise questions about the integrity of food. Overnight we realized that globaliza-

---

2. The Worker-Farmers is a farmers' movement set up in 1972 by Bernard Lambert from an opposition movement within the CNJA.

tion was forcing us to eat food that contained hormones. So on one side of the Atlantic a wholesome product like Roquefort was being surcharged, while on this side we were being forced to eat hormone-treated beef!

And some people were delighted that you took on an American symbol.

JB: Absolutely. But we quickly dealt with that. We didn't want McDonald's to be seen as the prime target. It's merely a symbol of economic imperialism. Besides, we never called for a boycott of McDonald's. The journalists grasped that pretty quickly, and most of them latched on to the ideas behind the McDonald's symbol. Our political leaders, however, tried to talk up the anti-American element: some by playing the 'typically' Gallic card, others by invoking 'sovereignty' in a way that fuelled nationalism. This was the populist side of things: it's easy enough to rubbish America, to discard a problem as not being of direct concern to us, rather than to confront it. From this point of view, it was very easy for our leaders to support our actions.

LIFE IN PRISON

And inside, how did you keep in touch? What was the talk about your actions behind the prison walls?

JB: All prisoners watch television. Six channels were available, at 200 francs a month. The guys in a cell often clubbed together to hire a television. Those who have been in for a while have a radio. Prisoners who got newspapers would often pass them from cell to cell. Given the TV coverage of the events, everyone knew the facts. A detained man spends twenty-one hours a day in his cell; this gives him ample time to surf all the channels. I had many discussions with the other prisoners. They're cut off from society, but even they understood what it was we were struggling against.

The prison warders would also come and chat with me. I was held under the same conditions as everyone else, but my presence aroused sympathy, because they knew I was inside for a trade-union offence. They would say: 'We've been demonstrating, too. So what's happened to you could just as easily have happened to us; we even burned down the prison gates.' They understood the risks of industrial action. In the mornings I would go to the lounge, where there was always a warder ready to hand me a newspaper – my own didn't come till midday. So we really struck up a rapport.

One night during my stay, a young eighteen-year-old traveller committed suicide. He'd been given six months without remission for stealing food. He wanted to be put in a cell with his cousin. In the time it took for this request to go through the hierarchy . . . he'd cracked up. This was the second suicide that summer in Villeneuve-lès-Maguelones.

How did you survive prison?

JB: The prison held seven hundred inmates. It was an intense experience. Life inside is very difficult, but there is a sense of human warmth. Petty criminal, murderer . . . or trade-union farmer, you all end up in the same boat. You end up alone, with your conscience, facing the acts that have led to your prison sentence. During the short spells of communal activities, your ability to get on with others is tested. Anyone who doesn't behave properly is soon marginalized.

Little by little, I got to talk with everybody, even including the 'hard guys' who'd been inside for a long time. The prisoner who'd notched up the most sentences had spent twenty-five years in prison, in three goes, and was awaiting a new trial. He was the most political of all, a real social rebel. He was forty-five and he'd been in the nick since he was twenty. He also had a number of escapes on his card. He was proud of never having done a day's work in prison, and considered me a comrade in struggle.

One thing became clear to me during all these hours of discussion: the irrationality of the penal system, with its distribution of sentences varying randomly from one law court to another. Sentencing is a lottery. The length of the sentence is often decided on the basis of no more than a person's appearance: the same offence can carry a five- or ten-year sentence, with no apparent reason for the variation.

Then there's rampant privatization, which is a real scandal. Apart from the warders, who are state employees, everything else is in the hands of the private sector. So you have to pay for everything: television, laundry, essential toiletries, and even meals if you want to eat properly. For those who are skint, there's the 'slop': first course, main dish and dessert – without salt, lukewarm, and so inedible it often gets chucked out of the window.

In fact, three times a week, prisoners are made to clear up the piles of rubbish that have accumulated between the cell walls and the railings in the courtyard. On the other hand, for the well-off, there are ready-prepared dishes made with fresh produce: duck breast, spring chicken, steak entrecôte. Some days you could buy cakes and gâteaux; brioches, croissants, pains au chocolat, tartlets and éclairs. The same outlet, often one of the big food chains, sells both the revolting stuff and the *de luxe* meals. I think the journalist Paul Amar, on one of his TV programmes, said that McDonald's had offered to finance some repairs in the prisons in exchange for being allowed to supply food to the inmates.

How did the other inmates relate to you?

JB: There was a restaurant owner from the Camargue, who had a whole load of problems: bankruptcy, family break-up, divorce. He had a real crisis, and – all alone, in broad daylight and unmasked – had attempted to rob a bank. Needless to say, he was arrested. He'd been in prison two years, and had returned to his favourite teenage activity, drawing and painting. He paints large canvases and frescoes in his cell. Canvases up to five metres high, made by sticking sheets of paper together. He gave the canvases to his sister, who opened an exhibition of them in a bistro in Sète. Two days before I was due to come out, he showed me a press cutting from *Midi Libre* with the headline: 'A prisoner of the Villeneuve-lès-Maguelones exhibits his work in Sète'. Below the article was another which read: 'José Bové refuses to leave prison'. We were on the same page! He offered me the cutting and some photos of the exhibition. On the Sunday after I came out of prison, we contacted his sister and arranged for the exhibition to come to the Cultural Centre in Millau for a week. The opening was attended by all the town bigwigs. With this guy's trial coming up soon, we were pleased to do it. In November 1999, he got thirteen years; painting is his salvation.

When I refused to pay the bail, I went back to my cell without telling the other inmates. They heard on the 8 p.m. TV news that I was back inside, and started tapping on the windows. It was all very moving. The following Sunday, friends from the Larzac and the Farmers' Confederation visited and held a solidarity picnic, with music and other entertainment. They drove round the prison hooting their horns. Those prisoners who were on the side of the building that looked out on to the access road and the bridge were burning paper by the window, hanging out sheets and shouting 'Free José!'.

When eventually I did decide to accept the offer of bail, thanks to a large support movement, the other prisoners expressed solidarity with my decision. Some were even prepared to give money to help with the struggle. On the Tuesday morning, before leaving, I first went into the courtyard to say goodbye to them. I didn't want to sneak off like a thief in the night.

And when you came out? What was the first thing you saw?

JB: I'd expected there to be a crowd, but not as many as turned up. I first saw Alice, my wife, and Hélène, my youngest daughter. Then François and the union friends. I barely had time to embrace Alice and Hélène, because there was such a large gathering. I told myself: 'I've got to brave this out.'

I'd planned my coming out from my cell and had worked out the statements I would make; but the moment I stepped outside, I no longer knew what I wanted to say. Here I was, surrounded by cameras and mikes, searching for words. After what I'd been through, I was afraid I was totally out of sync with life outside. It was a weird feeling.

I knew the external layout of the prison from watching television; I'd seen the press conferences at the doors, in particular on the day I'd refused to accept bail and returned to the cell. That had been very hard. The lawyer, Alice, my daughters, mates from the union – everyone, in fact – knew of my decision beforehand. Our lawyer, François Roux, held a press conference outside the prison gates and began by saying: 'José Bové refuses to come out.' I was on the other side of the wall, in front of a television, and I saw my daughter break down in tears. The camera showed a close-up of her, and I was badly shaken. Then the mike was taken by Raymond Fabrègues, the Aveyron spokesman for the Farmers' Confederation, and as he started to speak, he too began crying. I said to myself: 'What a start!', but at the same time it was very moving to see farmers in tears; it really touched me deep down. All these thoughts came back to me as I left prison.

Once you were free, you packed some things and went up north to Paris. . . .

JB: Well, in the first instance I went home; that same evening more than five hundred people gathered in the Larzac to celebrate. On 7 September, the five of us who had been convicted met up in Paris at the invitation of the support committee,[3] which was chaired by Henri Leclerc, president of the League of Human Rights. A press conference held at the Bourse brought together twenty to thirty journalists as well as representatives from all the

---

3. This committee brought together members of several major trade unions, including finance workers, and the union of French lawyers; the League of Human Rights; Greenpeace; rail workers, and a number of other organizations.

organizations who'd signed the petition for my freedom. These groups were all under the banner of the Committee for Citizens' Control of the WTO.

It was then that we discovered that our struggle had moved beyond a mere farming issue. The stakes had shifted to encompass opposition to the WTO, union repression, junk food – in short, globalization – and had succeeded in bringing together thousands of different voices. These people were to become the mainstay of the movement, which grew from strength to strength up and down the country, from the beginning of the school year until the glorious events in Seattle at the end of November.

The same day, a smaller gathering was arranged by the National Committee of the Farmers' Confederation, who warmly welcomed us into their headquarters. Meeting the people who had struggled for three weeks to obtain our freedom, and had kept up our fight, gave us renewed energy to continue. . . .

2

McDONALD'S:
DUFOUR'S STORY

Between the eclipse of the sun on 11 August and 15 August, Françoise and François Dufour were preparing for their escape to the Haute-Normandy, to spend a short working holiday checking out Deauville and its famous wooden boardwalk. The Farmers' Confederation had plans to intervene in the forthcoming American film festival to draw attention to hormone-treated beef and the retaliatory measures taken by the USA, notably the surcharge on French agricultural produce. François had been put in charge of preparing the ground. He wanted to install a farm on the boardwalk with calves, cows, pigs, chickens. But above all, he wanted to talk with the American film-makers – to tell them that if they chose to be culturally out on a limb, that was all very well, but the French farmers had no such choice, and did not want to become agriculturally isolated. It's a trap. We believe that our agriculture, with its land, its jobs, is worth defending. François had absolutely no idea that in a little while, whenever he spoke, the whole world would be listening.

'A MILLION FRANCS' WORTH OF DAMAGE'

As National Secretary and spokesperson of the Farmers' Confederation, were you aware at the time of what was being plotted in Millau?

François Dufour: Yes, like everyone on the National Committee, I knew. We'd been deliberating since spring over the conflict between the USA and the European Union on the issue of hormone-treated meat. The adoption of the Common Agricultural Policy (CAP) on 25 March 1999 in Berlin gave rise to a follow-up round of negotiations in the WTO, due to take place in Seattle at the end of November. The Farmers' Confederation approved unreservedly of the European Union's tough stance against the USA.

On 20 and 21 July we had a National Committee meeting to decide what to do about the Seattle negotiations. We discussed whether to go it alone or work with other groups. We examined the various possibilities: demonstrating in front of the US Embassy in Paris, or outside their tourist office; or maybe distributing leaflets to American tourists. From the very start, we emphasized the importance of not falling into the trap of anti-Americanism.

At the protests around the GATT meeting in Marrakesh in 1992, we had met a lot of people from the arts world. At two big conventions in the same year we condemned the market culture, the loss of cultural identity, the grip of market forces strangling our choices. In this light, the National Committee of the Farmers' Confederation decided to take advantage of the American Film Festival being held in Deauville in September, and expand our contacts with artists and others in the culture business, as well as win over public opinion. We wanted to explain to the American Festival-goers that it was not their culture we objected to: that it was very welcome in our regions, but that the multinational companies had to respect our differences, our identity. We don't want homones in our food; they're a risk to public health, and go against our farming ethics. At a more fundamental level, imposing hormones on us means that our freedom of choice in the food and culture we want is seriously restricted. Agricultural exchanges have existed for a long time: we don't advocate exempting agriculture from the policies of international trading, but we want something different from freedom of the market and the liberal economy.

I was put in charge of organizing this activity, which assumed national importance. In the middle of the meeting, the representative from the Aveyron announced: 'Roquefort is being surcharged, and a new McDonald's is under construction in our area.'

How did you react to the news of the arrest of the Aveyron militants?

FD: As I explained, I was preparing for Deauville when the news arrived from Millau. I was immediately contacted and given a full account that same evening by José. The first thing that cheered us up was the knowledge that there had been no violence. Whenever there's a day of action, the National Committee closely follows what happens and is keen to ensure that our actions and demands are understood and approved by public opinion, so violence, or the use of force or fear, is not part of our approach – unlike the FNSEA. On 12 August, we felt that the day had gone well – there had been a good turnout, no violence, and positive action. And though some legal comeback from McDonald's looked likely, this would allow a public debate to develop during the trial, as had happened with the anti-GMO actions.

José was the first to warn me on the Tuesday morning. My mobile is on the bedside table, and it rang at about seven. I recognized Bové's voice: 'The cops have been round to Raymond's, Christian's, Jacques Barthélémy's and Léon Maillé's houses, and are now looking for me. The guys were carted off; they've been taken to the cop station and it doesn't look good. We've got to keep a close eye on it.' I warned him that they would be after him, too, and he replied: 'They've already been round to my place; I wasn't there. I'm on holiday, but I'm not on the run. Let's stay in close touch.' I immediately notified our legal representative, Paul Bonhommeau, and our lawyer in Montpellier, François Roux. José kept in touch with us, as did the local representatives of the Farmers' Confederation, Gwenaël Latrouite and Christelle Combes, who called regularly.

Towards the end of the day, the radio news began to report that four farmers were in prison for having 'ransacked a McDonald's'. 'Destructive fever', 'A million francs' worth of damage' – these were the headlines which threw those of us who were in the Confederation into some doubt about what had happened.

I had full confidence in José and the Aveyron militants, but I couldn't help thinking: 'Could they have deceived us? What is this million francs' worth of damage about? If it's true that the building has been brought to the ground, how come no one told me about it?' I called the militants in Aveyron and asked them what they'd done to cause a million francs' worth of damage; they were astonished at the allegation, and challenged the sum. They told me that they had removed some tiles, torn some boards, twisted some drainpipes, which they had filled with sand and gravel. I got the same story from others I knew down there. There was no word yet about the warrant for Bové's arrest, but it was reported that he could not be located, with the local police claiming he was hiding in the Larzac. I said to myself: 'José phoned me, he's on holiday somewhere, he doesn't lie. So someone else must be lying.'

What was the reaction of the activists in the Farmers' Confederation?

FD: We were in the middle of the month of August; many of the Confederation staff were on holiday, so were the other national secretaries. At the Confederation's Head Office I met up with Benoît Ducasse and Paul Bonhommeau to try to alleviate their concern. They'd heard the radio and TV news reports of rampage, havoc, a million francs' worth of destruction, farmers imprisoned. They wanted an explanation of what had happened. So did the media.

The following day, I asked everyone to be ready to take action, because there were farmers in prison and the judge seemed to be taking a tough stand. I was still getting a lot of negative feedback of the sort: 'But listen, we're dealing with a million francs' worth of damage here. Didn't you hear it on the radio this morning? It's a ransack!'

So I composed a letter and repeated what I'd explained to everyone over the phone. I wanted to get it down in black and white, clearly, that the alleged devastation was not a ransack but a mere dismantling, that we were confident in stating this, and that we had to stop beating about the bush. Our task was to take on board the judge's severe attitude, and realize that we were dealing with union repression rather than debating the cost of the damage. In the end, the Seine and Marne Confederation, which had been the first to declare: 'Although we're all in favour of an occupation of McDonald's and denounce McDonald's and the judge's strategy, we can't accept a thousand grands' worth of damage', were also the first to go down to Millau. The others soon followed.

What happened then?

FD: With our lawyers, François Roux and Marie-Christine Etelin, we followed developments as closely as possible. From the way the judge handled the case, we became increasingly aware, from one hour to the next, of the extent of her anti-union bias. She evidently wanted to isolate us within the farmers' union movement.

Eight days after the action, I went with José when he turned himself in at the court of justice. We'd organized a demonstration, in which I took part. I went to visit the McDonald's for myself, and was pleased to see that the workers were putting the finishing touches to it. I said to myself: 'There's no way now they can pull the wool over my eyes.' The estimate of the damage caused was due that very same day. One hour before their appointment, the assessors from McDonald's insurance company said:

'Gentlemen, we've got to put off the meeting till tomorrow, or the day after tomorrow. We'll let you know.' Since then, not a word, and there's never been an assessment.

On the strength of what I saw of the 'ransack' at McDonald's, I confronted some of the newspapers which had carried the lurid reports – notably *La Dépêche du Midi* and *Midi Libre*: 'You were the first to mention a million francs; who gave you that figure?' One journalist replied: 'The managing director of McDonald's announced it on the evening of the action.' I then called the Aveyron prefecture, which had reported that 400,000 francs' worth of damage had been done, and asked them how they'd got to that sum. I then asked to see the assessors' report: of course, no one could produce it; it didn't exist. When we met some of the McDonald's site workers, they confided to us: 'There was 30 or 40,000 francs' worth of damage. There's been no delay in the rebuilding and the McDonald's will open, as scheduled, on 21 September.' On the strength of this, we insisted that the militants clearly explain that the dismantling had been a symbolic protest against multinationals like McDonald's taking over the world.

Judge Marty's 'One law for them, another for us' attitude really made us angry; We questioned why this supposedly independent justice system, so acclaimed by the government, had never prosecuted the activists who had wrecked the Ministry of the Environment offices[1] in Paris, nor the violent Breton pork producers on the FNSEA demonstrations. It seemed paradoxical to us that the powers that be had chosen to defend multinationals like McDonald's and the manufacturers of hormones and GMOs, against farmers and consumers.

---

1. In August 1998, about a hundred agricultural workers, sent by the union's Paris leaders, had ransacked the offices of Dominique Voynet, Minister for the Environment.

A 'FARM OF THE FUTURE' AT DEAUVILLE

Did the media coverage of the Millau action thwart your plans for Deauville?

FD: I was forced to juggle between taking charge of the McDonald's crisis and continuing with the preparations for the actions at Deauville, scheduled for Sunday 14 September. I had to give assurances to Madame Irnano, the mayor of Deauville, the Prefect of the Calvados area, and the Director of the International Centre in Deauville that we were not men of violence: that we would not wreck the International Centre, nor the local McDonald's. I vouched for all the farmers, and the authorities agreed we could go ahead with the demonstrations as planned. We needed to get more attention for the Millau actions, and show that we wanted good relations with the Americans.

On the day itself, 350 farmers were involved in setting up a farm on the boardwalk, just a few steps from the village and the Festival: cows, horses, sheep, ducks, turkeys, chickens and rabbits were put on the straw which we'd carefully laid out. The local town council, the Festival organizers, passers-by – indeed, everyone – welcomed it with open arms.

The Festival's president, Régis Wargnier, had not forgotten how, in 1993–94, we'd supported his campaign to allow French culture to remain outside the GATT agreements, and came to greet us to show support for our action. We created a festive atmosphere, with a farmers' market, banners, singsongs and grilled sausages, and our 'farm of the future' achieved its objective: dialogue between farmers and the thousands of townspeople who came to meet them. The local Basse-Normandy group, the CGT, the CFDT, political activists on the left as well as the right, all offered their support and participated in drafting the outline of a united front committee for citizen control over the WTO. The Millau action had started the ball rolling, and the clamour for José's release testified to the sympathy growing around us.

Did anything come of the meeting with the film producers?

FD: After Deauville, a group of film directors came to the Larzac. A TV programme was broadcast live from the Maison du Larzac at Asse. The film directors – including Pascal Thomas, Robert Guédiguian, Gérard Guérin, and Jean-Henri Roger – all shared our views on globalization, and reaffirmed their opposition to any attempt to trade agricultural with cultural products at the WTO negotiations.

In your opinion, what caused such a widespread movement of sympathy for José Bové and the Millau action?

FD: The bulk of the sympathy resulted from the fact that we had consistently explained that the struggle we were waging was against the downgrading of agriculture, against the WTO. Over the past three years, consumers have been exposed to a number of health scares – benzodioxin, mad cow disease, swine fever, to name but a few – and they've had enough! Town-dwellers understand that an attack on the countryside and the quality of its produce is an attack on the relationship between the farmer, his land and the consumer. It is precisely this relationship which is missing in produce affected by food scares.

The support movement was evidence that people understood the ideas we had been spreading for years: that agriculture is not a separate sector, that it can't be reduced to just another aspect of production. Eating habits, quality, gastronomy, cultural identity and social relations all depend on farming, and define what we refer to as agriculture. It follows from this that the farmer's fate is indissolubly linked to that of all other citizens.

In the farming community, the demonstrations of support went well beyond the confines of union affiliation. The only person who didn't seem to get the point was Luc Guyau, President of the FNSEA, whose sole contribution was to vilify Bové and denigrate the action in Millau as a typically Gallic stunt. Clearly he saw the threat to his own position of a mass movement legitimizing the action, and hence the Farmers' Confederation. That's why he tried to discredit Bové. But he was marginalized. Many people identified with the action, just as had happened when we had taken strike action in support of the railworkers in December 1995.

Many voters shunned the latest European elections, because there was nothing for them to relate to. People are not against the building of Europe *per se*, they simply want to protect the traditions and rhythms of farming life. Many abstentions and votes cast for the candidates on the 'Hunting, Fishing, Nature and Tradition' (CPNT) slate result from this loss of identity. It wasn't just the hunters who voted for the CPNT; there were many who felt a loss of cultural identity.

Agricultural identity is part of this; you don't have to be a farmer or live in the country to feel rooted in the land. Such roots connect all parts of the country in a unifying whole, and this can't be undermined by Europe or globalization. The McDonald's issue came just at the right time to stir up such feelings. Even the most liberal economic milieux had to admit that the downgrading of agriculture and its appropriation by factories was destroying those roots. People don't want to be uprooted. This is essentially what public opinion boiled down to – much more than to a sense of solidarity with the economic hardship being suffered by the producers of Roquefort and other affected products.

TRADE-UNION REPRESSION

For the first time in this country, the law attached the conditional release of the imprisoned farmers to the payment of bail for each one. Were you prepared to pay?

FD: From the moment the first four comrades were imprisoned, we called for their release at the court in Montpellier. This was on Friday 20 August at 11 a.m. Our lawyer, François Roux, announced: 'It looks like we'll get bail at 100,000 francs each.' Oddly enough, this figure, 400,000 francs, was the same as the one quoted by the prefecture for the damage caused to the McDonald's. When the verdict came, bail was set at 105,000 francs per prisoner, so 420,000 francs had to be found. At 2 p.m. it was José's turn. The judge refused to set him free, even on bail, and put off the next hearing to a fortnight later. The guaranteed cheque for the others' bail had to be deposited at the tribunal before four that afternoon.

Bail, in our eyes, was tantamount to trade-union repression. It's a means of destroying a union organization, bringing it to its knees. For an organization like ours, the sum was enormous. We were aware that by paying the bail we were setting a precedent, contributing to the establishment of an Americanized two-tier justice system in France: those who can afford it can leave pre-trial prison; those who can't have to rot in jail.

When Bové was eventually granted bail, were you prepared to pay it?

FD: José had let it be known that he was not willing to buy his freedom. We respected his wishes, though there were many militants who would have coughed up for him. The Roquefort Association offered to pay the bail. Respecting José's wishes, we agreed that it was only the mass movement putting their hands in their pockets that would liberate him. We suggested to the Association that they should donate the money to the Millau support committee which had just been set up. They refused, making it obvious that their concern was more with the image of Roquefort than with any commitment to building solidarity.

José's refusal initiated a debate on trade-union repression and the arrogant attitude taken by the law. The leaders of the winter 1995 movement decided to set up a support committee that very quickly attracted national and international support.

What news were you getting from Bové at that time?

FD: I was in daily contact with our lawyers, who visited him every day. José was psyching himself up to spend time inside; he's tough enough to withstand this, but that didn't stop me worrying about the length of his confinement. Ten days previously, I'd met the wives of Raymond Fabrèques and Christian Roqueirol — very unhappy women — and I was thinking of José's partner, Alice, whom I couldn't reach. How long was her husband going to stay in jail? I no longer knew what to do. Was his family prepared to endure this ordeal? It was also his partner's choice. We asked ourselves these questions, and found no easy answers. We in the Farmers' Confederation weren't used to having our militants imprisoned. At most, we'd had some nights in prison following land struggles at the time of the CNSTP.[2]

So it was the first time one of your militants spent a considerable time in prison?

FD: Nineteen nights — yes, it was the first time. We'd had a few trials in the past for actions which had annoyed the authorities. We'd been lucky not to have had people sentenced, but several were arrested for just causes which the public understood. The law takes such things into consideration — something the judge in Millau failed to comprehend when she ordered the search for further militants. Even when the President of the Republic, on tour in Quebec, had declared: 'On this issue we will not make concessions' when he was questioned about hormone-treated beef and GMOs. And the Prime Minister, on tour in Corsica, had stated that he considered the cause of the Millau farmers to be just. In the face of such declarations, which were lending legitimacy to our action, the judge should have realized the extent of public support. She could have proceeded with more intelligence.

2. The National Confederation of Farm Workers' Unions, created in 1981 on the initiative of the Worker-Farmers, brought together a number of unions and local groups which broke away from the FNSEA (see Chapter 3).

SOLIDARITY FROM ALL QUARTERS

There were numerous declarations of support for your action and for the release of Bové, representing a broad range of political and trade-union opinion. Can you describe this momentum?

FD: In addition to the support we'd expected from activists, there were many who rallied behind us from unexpected quarters, including lemonade merchants who refused to sell Coca-Cola. The majority of French people, from royalists to left-wingers, discovered a new form of trade unionism in the agricultural sphere; one that denounced inequalities and struggled for work and for a redistribution of public funds, and had an international outlook. The Millau action was like a spark that ignited a new debate. While the social crisis worsens, the political leadership has totally failed to provide any debate. We were concentrating on the issues of food – everyone holds their freedom of choice in this matter dear, and expects effective food safety controls – but public concern also extends to health, education and culture. The protests at the WTO Seattle Conference brought a lot of these concerns together. Add to this local negotiations on the thirty-five-hour week and ensuing discussions on the right to work, and a general upsurge seemed to be in the air. The farmers' movement could not fail to be affected by it.

When Bové declared: 'You can't buy trade-union freedom,' we were inundated with phone calls, including calls from the USA, all saying: 'José will come out of prison, because we'll pay the bail.' Cheques for 50, 100 and 500 francs poured into the support committee.

During the weekend of 28 and 29 August, the phone in my house never stopped ringing, with farmers and union militants inquiring how much was still needed, where cheques should be sent. I couldn't do any work on the farm, and although it was time-consuming, I realized that a considerable momentum was building up.

The Farmers' Confederation is a member of the CPE and of Via campesina.[3] *What support did you receive from these organizations, and from the international movement generally?*

FD: Letters came in from all over the world – by post and by Internet. The first to arrive was from the American farmers of the 'National Family Farm Coalition', who are fighting the industrial model in the USA. They belong to a movement 'For a Sustainable Development', close in spirit to our own 'Sustainable Farming' coalition. The American consumers and farmers between them sent us more than 30,000 francs.

Since the Millau affair started, we've been able to communicate quickly and easily with the whole planet thanks to the Internet (our federation was the first French farmers' organization to have its own website,[4] which we've used extensively, especially to exchange information about GMOs). We also received support from *Via campesina* through our representative at its meeting in Africa.

In the end, our campaign was covered by TV, radio and newspapers everywhere: throughout the Anglo-Saxon, European, Asian and South American world. We had to find English, Spanish and Portuguese speakers for interviews. In the corridors of our offices we would constantly run into journalists. Some days it was difficult to cope with it all. American reporters from *Newsweek*, the *International Herald Tribune*, *Seattle Times*, *Washington Post*, ABC, CNN, NBC – all wanting to know who we were and what we wanted. They were quick to make the link with the anti-GMO movement in Europe. We also met Brazilian political and economic leaders to develop a supply line of non-GM soya beans from France.

3. *Via campesina* (Farmers' Road) is an international farmers' movement grouping together 69 farming associations from 37 countries in four continents.
4. www.confederationpaysanne.fr

3

JOSÉ AND FRANÇOIS:
THIRTY YEARS PREPARING
THE GROUND

François and José are the same age. François is from the Manche area of Normandy, and is the son and grandson of farmers. He now runs the family farm with his wife. José, who has roots in Bordeaux, is the son of researchers in the food industry and the grandson of market gardeners; he is a member of an agricultural co-operative, GAEC.[1] The two are really four – without their partners, Françoise and Alice, the two men would not be who they are. François made a detour into intensive farming in the 1970s, before being converted to sustainable agriculture. José had always dreamt of organic farming, but settled down to 'mainstream agriculture' without ever being tempted by technology or its pitfalls.

They are separated by the river Loire, but united by an interest in milk production: for François, it is cow's milk; for José, ewe's. Both have been militants for a long time, interested in collectivism since they were young.

---

1. GAEC (Groupement agricole d'exploitation en commun), established in 1962, is an agriculture co-operative enabling several farmers to work together.

When François got involved in a struggle defending his job, José was already fighting to change the world. To his own cost, François found out about trade-union sell-outs; José had always been on the anti-authority side. François had followed the classic path of activism inside the structures of the FDSEA, where he had a promising future but renounced it all when he rejected the two-faced talk of the leadership – one for the membership, the other for the Minister. Between violent ultra-leftism and fireside theories, José stuck to his militant line of redistributing wealth while never resorting to violence.

The fact that these two men remained true to their beliefs brought them together in a movement called the 'Worker-Farmers', set up by Bernard Lambert within the CNJA, which became an autonomous movement in the aftermath of May 1968. In 1987, it became the Farmers' Confederation.

Those who know François Dufour and José Bové agree that they have one thing in common: in the midst of a struggle – for example, when tension mounts at the head of a demonstration, or as a rally dissipates, when there's wavering or hasty reactions – they both know how, where or when to take the initiative. It's a sort of instinct that simultaneously anticipates both the membership's desires and the opponent's reaction. Both have a good nose. They don't need lengthy political analyses to know what's possible. They can immediately evaluate the political potential of an action. They can sense whether the membership is willing to go further, realizing that without them there can be little success.

This was how the march from the Aveyron to Paris was organized; how a farm was set up on the boardwalk at Deauville; and even how a McDonald's got dismantled.

Here is a quick overview of the careers of François and José, friends and accomplices: two different men, who came into the limelight together as a result of their convictions.

JOSÉ: TRAINING IN THE LARZAC

The first thing you notice about José are his eyes – a luminous blue, full of warmth. His smile is never far away – the lines around his eyes always seem to herald it. For the people of the Larzac, José Bové is the farmer with a hat – a mountain hat, to keep his head warm in the cold and windy weather of the plateau. It is a hat that smokes: Bové rarely ventures out without several pipes, tobacco and other utensils, all kept in a leather satchel. Both libertarian and pragmatic, he has a taste for subversive, symbolic actions, and derives strength and a certain contentment from his beliefs.

Nothing much surprises Bové, and if he has doubts he doesn't show them. He keeps his private life to himself and, for a rebel, is very discreet. He neither hunts nor fishes, but he reads a lot. He has acquired a fine knowledge of the international workers' movement and the history of peasant revolts. Henry David Thoreau remains one of his favourite bedside authors. He admits to a love of the sea, which he shares with his partner, Alice. Every year without fail they leave their farm in the Larzac behind and head for the Atlantic, to a little spot on the coast of the Landes, where blue and dark green mingle right up to the horizon. That's where they had escaped to for their summer break in 1999, after the dismantling of the McDonald's in Millau. That's also where they learned that the police were looking for José.

José Bové has spent his life reconciling his actions with his beliefs. He was barely fifteen in 1968, when he began to challenge the school system. The following year, the focus of his criticism was society itself, which he considered too militarized and too much under the nuclear shadow. These concerns kept him at a distance from left-wing groups, and he got involved in activities such as flyposting with the militants of the disarmament movement, campaigning for peace and freedom and defending conscientious objectors who, in those days, faced a life sentence if the military tribunals had their way.

Armed with his Baccalauréat, Bové enrolled for a philosophy course at

Bordeaux, his family's home town. He barely set foot on campus, preferring anti-militarist activism to meditations on Plato. His convictions left him no choice but to refuse military service, and he sought recognition as a conscientious objector.

One day in 1971, as he was painting an anti-militarist banner, he began to get to know his co-painter, Alice Monier, an undergraduate in political science, a year younger. Their love story began there, 'with a banner which drew 1,500 people behind it through the streets'. The two lovers hung around networks of young Catholics and non-violent libertarians. Here, José developed his thoughts on the role of political action and leadership. He rejected the idea of power coming from the barrel of a gun: 'It's always the people who end up being shot. We certainly don't want total freedom for those who are armed to the teeth.' Jacques Ellul, professor and theologian in the Social Sciences Department at Bordeaux University, provided José with a theoretical understanding of the role of the state and the technology which structures it: 'He was the first to theorize the autonomy of technology, to show how both the state and the economy are creatures of a technology which has its own inner logic.' This understanding brought José and his comrades to a different radicalism from that of the far-left groups of the time. These groups were locked in debates about the logistics of seizing power, whereas José was challenging the intensification of farming. He was also one of the very few who volunteered to work, every Friday, at the military tribunals, aiding those who were detained and keeping files on the proceedings – a mammoth task which turned out to be extremely useful to the lawyers and militant young magistrates opposing the trials.

The local anti-military network was one of the first to mobilize against the extension of the military camp in the Larzac. Unsure how best to take forward a campaign of farmers opposed to the army, José and his friends set up a Committee for the Larzac. Their campaign was assisted by the stance of Lanza del Vasto, an apostle of non-violence and spiritual leader of the Community of the Arche, nestling in the foothills of the Larzac. His fifteen-day fast at Easter 1972 galvanized the farmers on the plateau, and they vowed never to sell to the army and never to leave their land, regardless of of what they were offered. Their oath was reinforced by national support. The first demonstration, in support of a hundred and three farmers threatened with eviction, took place outside the Larzac, in Rodez, on 14 July 1972. José's Bordeaux committee helped to steward the demonstration, which was committed to non-violence. From that point on, the Larzac became the prime organizational focus for Bové and his partner.

At a meeting in Hospitalet du-Larzac, at Easter 1973, Alice and José met Bernard Lambert, leader of the Worker-Farmers organization, who suggested that a rally should be held in the summer for all French farmers on the Larzac plateau. José remembers Lambert, who died in 1984 at the age of fifty-three, pacing up and down the aisle in a parish church as he harangued the farmers, who were sitting quietly in rows: 'The rally was Bernard's idea. He hadn't discussed it with the Worker-Farmers; it was one of his political intuitions. When he felt the need for something, he threw himself headlong into it.' José admits acting in a similar sort of way himself; this sometimes creates misunderstandings about – even hostility towards – his union activities. But Bové's intuition, like Lambert's, has generally proved reliable.

The rallies in the Larzac in the summers of 1973 and 1974 saw the coming together of a wide range of different groups: angry farmers, strikers from the occupied Lip watch factory, conscientious objectors, protesting prisoners, feminists, local militants and a good section of the far left, including the radical wing of the French Socialist Party. Alice, José and the Worker-Farmers participated in the construction of a sheep barn at La Blaquière, an action which came to symbolize the Larzac resistance to the encroachments of the army. Lively debates took place between the Worker-Farmers from the west of the country, who were very heavily involved in industrialized agriculture, and Alice and José, who, with their opposition to technology, favoured a form of sustainable farming. Another topic of discussion was the relationship between farm tenants and landowners.

Bernard Lambert answered the question with his customary clarity, declaring that the conflict was not between the poor farmer and the rich farmer who exploited agricultural labourers, but between all the farmers on the plateau and the army – in other words, the state.

SUBVERSION IN A STATE OF NATURE

It was at around this time that Alice and José decided to move to the countryside to work on an alternative agricultural project. They chose to settle in the Pyrenees, near Béarn, where they struck up a friendship with Jean Pitreau, a farmer who took on volunteers and conscientious objectors to work on a mountain farming scheme: 'It was a brilliant idea, outside the confines of the FNSEA. It consisted of showing farmers who were not part of the system that the work they were engaged in was both valuable and feasible.' Alice and José signed up as volunteers, and spent the next three months taking their first steps in farming as agricultural labourers, high up in the Aspe Valley, part of the Basque Country. Cows were milked manually, and the milk was made into cheese. All this stood them in good stead later on. Although it was short-lived, this experience forced them to test their theories of organic farming in practice.

September 1974: Bové, twenty-one years old, was refused the status of conscientious objector. For almost a year, trying to avoid prison while his appeal was pending, he hid out on a farm, where he learned to make butter and yogurt from cow's milk. During this time, Alice was working part-time in the archive department of the daily *Sud-Ouest* newspaper. The couple lived discreetly in a friend's house. In October 1975, a few days after the birth of their first daughter, Marie, they heard that Bové's appeal had been successful. But on their return from registering the birth at the Town Hall, they found their house surrounded by police, and Bové was arrested. He let himself be taken to the police station, then he produced the letter from the State Council which he had received that very same morning.

During that year of 'hide-and-seek' and furtive meetings, José and Alice continued their trips to the Larzac, where they contributed to campaigns built around slogans such as 'Sheep, not guns' and 'Wheat gives life, weapons kill'. Not knowing whether they would be allowed to settle in the region, the couple began travelling back and forth between the Larzac and the Pyrenees. They wanted to live on the Larzac plateau, because it brought together the two things which motivated them most: the desire to live on a farm, and the anti-military struggle.

In the winter of 1975, the Larzac farmers decided to squat farms that had been bought by the army, so that they didn't stand empty. The initiative was carefully planned, and advice was sought from the geography specialists on the support committee. Alice and José volunteered to be squatters, and moved in at Montredon, on the edge of the military camp, in February 1976. They still live in the same house today, but at that time there was no road, no water, no electricity or phone.

Since the First World War, Montredon had been deserted, with only shepherds passing through. The village was almost entirely owned by the army. A crafty speculator, who had been tipped off by an MP friend about the plans for extending the military camp, had bought the whole lot up – together with a further 1,000 hectares and two farms – for 350,000 francs. He then sold everything back to the army for 45 million francs. Only two houses were left out of the deal: a second home whose owner refused to sell to the state, and a beautiful mansion, bought in 1975 by Larzac Universities, an association of teachers and academics from all over France who wanted to stand shoulder to shoulder with the farmers, against the army.

Alice and José chose a building belonging to the army, and started a survival farming project. It was tough going, as no bank was going to lend money to squatters on army property. But thanks to a helping hand from a farmer in the next valley who was willing to say that they were renting land from him, they were able to declare themselves farmers and claim social security. The same farmer offered to rear a hundred ewes for their meat, and a tractor was bought with the help of the Larzac committee.

So this was resistance farming, supplemented by a little building work which José managed to pick up locally. Life was hard, with daily harassment from the military. Water came from rain collected in a cistern, lighting from petrol lamps; the roads were virtually impassable, and money was scarce. In case of an emergency, such as clashes with the army, they were able to contact the outside world thanks to an ingenious bit of DIY. One night they laid a hidden telephone cable, cross-country, so that they could contact the nearest farm next door to a payphone. A real field telephone!

In June 1976, José joined some twenty others who broke into La Cavalerie military camp and made off with documents regarding the sale of land. Three weeks in prison followed. Then, in 1977, he was to be found behind the wheel of one of the ninety tractors which drove on to the military firing range, each with a dissident soldier on its mudguard, delegated there by soldier committees. Subversion on the plateau! 'It was surprising,' remembers José. 'You could always push the subversion one step further, presenting it as a logical outcome. That's still what's happening today.'

These were the years of courage, the time when close links were forged between all the players in the struggle – links which explain the strength of the reaction to the present-day surtax on Roquefort. This was the time when the bonds between the new arrivals in the countryside and the established locals became so close that they did not disintegrate after the struggle ended. The new arrivals brought along in their rucksacks – together with their determination and their strong right arm – their dreams of changing the world; the locals offered to share their most precious possession, the one which wears them down and bends their backs: the land of the Larzac. This land is not a mere backdrop, nor a romantic symbol for townspeople, but a source of conflict between the pleasures it provides and the hard toil it imposes on those who make a living from it.

José and Alice settled down and put flesh on their dreams. They made no attempt to hide the fact that they were newcomers to the countryside; on the contrary, they used their difference as the basis for a dialogue with the native farming community, forming a mutual solidarity which is still strong today. If, twenty years along the line, the sons of farmers now return to the land with a degree in their pockets, they will be accepted more easily, because a taboo has been broken by José and his partner, who combined educational achievement with considerable effectiveness as farmers.

In 1978, their daughter Hélène was born. It was the first home birth on the Larzac plateau, but certainly not the last. Alice and José belonged to a meditation group, including a doctor and a midwife, which opposed increasing medical intervention at birth. Until 1990, this group was the largest in France fighting for home delivery.

At about this time, the main farmers' leaders in the Larzac were delegates to the FDSEA, whose president was Raymond Lacombe. The union generally supported major initiatives in the Larzac, but tension arose between the orthodox supporters of the FDSEA and numerous support groups, who were often much more radical and, in some cases, included large numbers of members who belonged to the Worker-Farmers.

The young squatters of the Larzac were not attached to a trade union until 1978 when they set up an Area Centre for Young Farmers (CCJA). Alice soon became its vice-president, negotiating for a militant position of 'not a square metre to the army' while keeping links with the Worker-Farmers.

In 1979, José and Alice exchanged their flock for milking ewes, and began to turn the milk into cheeses which were then sold at local markets: an unheard-of activity on the plateau, where their companions in struggle earned their living from selling milk to the big Roquefort dairies. Adapting to their precarious financial situation involved running the milking machine off a generator, and ventilating the cellars where the cheeses matured with an old fan stripped out of a Peugeot 203 plugged into a 12-volt battery.

One day, Michel Fontaine, who was then professor at the Lyon veterinary school, came down to the Larzac with a group of students. This visit gave birth to the Vets' and Farmers' Association, which is still going strong today.

At a CDJA conference in February 1981, without any prior consultation, the main farmers' unions agreed to accept a small extension of the army camp. After expressing forceful opposition, Alice and the other CCJA representatives walked out and resigned.

THE LARZAC: A LABORATORY

On 3 June 1981, President Mitterrand kept his election promise and cancelled the extension to the camp. For those living on the plateau, the real bonus lay in the 6,300 hectares which were not bought back by their previous owners, and thus became state property. In the meantime, the Worker-Farmers had become the National Union of Farm Workers (CNSTP), and urged the CDJA members who had resigned to join them – this they did in the autumn. Working with the office of Edith Cresson, then Minister of Agriculture, José and the CNSTP militants deliberated on how to use the new land. Under pressure from the FNSEA, the government rejected an idea for common land with no formal ownership, but José and his friends maintained the spirit of this idea in their talks. 'This stock of geographically close plots of land, free from any private ownership, was an unprecedented opportunity for agricultural innovation,' José recalls, 'but a legal solution had to be found for its collective management.' This was achieved by setting up the SCTL, a collective which entered into a long-term lease with the state – a highly successful model which subsequently attracted the attention of other collective projects.

In 1983, for the tenth anniversary of the first Larzac gathering, the CNSTP held their Annual Conference in Millau, on the theme 'The Farmer's Status'. This meeting of the plateau inhabitants and the trade-union movement included a memorable banquet and ball held at the Blaquière sheep barn.

It was during this period that José met François for the first time, at a national meeting of the CNSTP in Paris. José doesn't remember exactly when – he feels he's always known François: 'In the struggles over milk and landownership, François was always there. We met around those issues – he and his people struggling in the Manche, never missing a chance to stick the boot in, and us, doing more or less the same thing in the Aveyron – we would meet up in Paris to discuss the situation.' They became – and remain – friends and accomplices. They understand each other in half-spoken sentences. If you'd seen them together in the streets of Seattle in autumn 1999, leaving their banner with friends to disappear off amid the haze of tear gas to go and check the US security forces close up, you'd understand their camaraderie.

Electricity arrived in Montredon in 1983, the telephone in 1984, and finally, in 1987, running water. Today, only one farm on the plateau remains empty. The farmers proved that by concentrating on working together, they could become economically viable.

In 1985, Alice left work on the farm to set up and run a centre for rural innovation. With the participation of farmers, biologists, sheep-breeders and others, this developed into a region-wide organization known as CIVAM.

At the National Conference of the CNSTP and FNSP in March 1987, José demonstrated how Roquefort cheese production was an example of the drift towards intensive agriculture. In the same year, the Farmers' Confederation was set up, with José as a member of its national secretariat, a post he held until 1991. During these four years he attempted to build a genuine alternative to the FNSEA. He remained on the Farmers' Confederation National Committee until 1997, when, in compliance with the Confederation's rules, he stepped down. Fixed-term elections, unusual in mainstream trade unionism, are part of the Farmers' Confederation rulebook.

Back in the Aveyron, Bové reorganized the production of cheese with a more equitable division of labour, and established a joint project with neighbouring farms. Their stock included sheep, cows and pigs for meat sent straight to the local markets. Half the ewe's milk was sold off to the Roquefort producers, with the remainder used to make homemade cheeses.

With any profits that accrued, the partners preferred to create new jobs rather than accumulate capital. Bové's daughters, Marie and Hélène, are students at Bordeaux University – one of History, the other of Law – and have no ambition to pursue a career in agriculture. In any case, inheriting farmland is forbidden. But José, pipe clenched tightly in the corner of his mouth, remains convinced that there are still changes worth fighting for.

FRANÇOIS: A CARING MAN

François Dufour represents the third generation of his family who have farmed at La Binolais in the canton of Saint-James, close to Mont Saint-Michel. His father bred horses: some twenty French breeds on twenty-four hectares. From the age of seven François enjoyed riding, and was far more interested in show-jumping competitions than in looking after the ten or so cows which provided much of the family's income. Until 1970, his parents churned and creamed the milk themselves. François complemented his farming childhood in Normandy with a school certificate in professional agriculture and management training.

François's father was a member of the FDSEA, and his son noticed that access to land for small farmers was dependent on the cost of farm leases. The right to these leases had been won in 1946, allowing farmers to exploit the land for the duration of the lease. Since then, prices had been pushed up by the bidding of solicitors, cattle dealers and local landowners; prices that were within the reach of the big farms, which continued to expand, were pushing out the smaller farmer. This was a taboo subject at the FDSEA, which pretended that the needs of the small farmers were unimportant, as small farms were 'unprofitable and fit for the scrapheap'.

The problem nagged at François, and in the early 1970s he started 'to look for others who thought as I did: that things couldn't go on as they were'. He joined other like-minded small farmers facing similar tenancy struggles.

In 1974, this group hit the road to the Larzac for the summer gathering there. François arrived 'anxious about being involved in direct action, but also hopeful of finding the necessary spirit to fortify the movement back home'. The Lip workers, in their overalls, impressed him enormously, as did the wide range of other social movements in the area. François discovered the effectiveness of workers joining together in struggle – something he had hitherto been a little frightened of. He remembers Mitterrand being heckled by the left, and above all a meeting with the Breton farmers – the beginning of what turned out to be an important relationship.

After the intensity of his days in the Larzac, François felt a little down when he returned to the family farm, anxious that he was not strong enough to sustain the sort of activity he had just witnessed. For a while he was miserable, but once he realized that he hadn't gone alone to the Larzac, and that there were others he could work with, he began to see the world in a different light, and became much more conscious of trade unionists' struggles against their bosses, and especially the conditions of the workers employed by the farm co-operatives: 'Living and working conditions, trade-union recognition and status – these problems were no different from ours.'

It didn't take François long to see the limitations of the local FDSEA union: at its meetings, the more radical proposals were censored, and little debate was allowed.

François met his future wife, Françoise Bernier, at a JAC meeting, and in June 1976 the couple took over the running of the Dufour farm. Like her husband, Françoise had grown up with horses and cows, but she had escaped life on the farm to work in an architect's office. Running a farm is not a piece of cake. Caught between the pressures of agricultural technology and the demands of the bank, the couple drifted into the very form of farming that they'd previously rejected. Forgetting what they had learnt at the Larzac and the JAC, they began intensive farming. If they wanted to claim the allowance available to young farmers, they had no option but to sell their horses and produce only milk. That meant buying extra cattle.

The plan was to have up to forty cows. This required an additional six hectares, and building new barns. These concrete sheds, where the cattle can feed themselves, are supposed to be a big time-saver, but instead, François had to double the hours he spent sitting on a tractor. He ploughed up the natural meadows, mowed up the white, red and violet clover fields, harvested the beetroot and cabbages for fodder. In doing so, though he didn't realize it fully at the time, he was destroying all the ecologically sound farming of two generations of Dufours who had planted fields with maize and ray grass.

Their first summer in charge of the farm saw one of the worst droughts the country had ever experienced – a disaster for the fodder crops. From September onwards, François and Françoise had to buy animal feed but had difficulty, at first, in judging the appropriate quantities. 'We just produced as much milk as possible, regardless of cost.' As soon as the money from sales reached their bank, it was quickly redirected to repay loans and to purchase feed and fertilizer.

As well as milking the cows, Françoise took care of the children who arrived to brighten up their life: Sebastien, born in 1977, and Nicolas, born the following year.

REJECTING INTENSIVE FARMING

The introduction of the 'joint responsibility tax' in 1978 was the last straw. A Departmental secretary of the FDSEA arrived to try to win support for this legislation from Brussels. The tax meant that small milk producers were expected to pay for the overproduction of large ones. The FDSEA official suggested that to oppose this tax was tantamount to blocking the development of Europe. 'At this point we stopped having anything more to do with the union,' explains François. He left the local FDSEA, set up a 'solidarity committee of small and medium-sized farmers' with his friends, and travelled all over the Department with a petition entitled: 'Joint responsibility tax: The FDSEA is betraying you'. They collected seven hundred signatures, but then did not know what to do with them. So they handed them in to the Prefect, who filed them away.

After this, in the name of the committee, François organized support for farmers in their disputes with landowners and bureaucrats. The media coverage of these confrontations attracted new supporters to the struggle. The committee was by no means the only one of its kind in France, and its formation was symptomatic of the discontent within the FNSEA. At this time, François was active in assembling rallies, organizing solidarity picnics, helping with discreet distributions of leaflets outside factory gates – all to aid the struggle. To dispel the corporatist image of the traditional farmer, he took time out to bring food to the picket lines outside the textile factories in the area, and to the postal workers, who were also on strike. He brought farmers along to talk to the workers, and to help to develop cross-sector links.

Meanwhile, François ploughed, sowed, and fertilized the soil heavily from morning till night: 'On the same patch of land we grew maize intensively, with a very high yield, and ray grass from Italy. We'd smother the ground in nitrogen, to give the maize a kickstart, then cover it with more nitrogen so that we could get grass in the spring. We never questioned what we were doing to the underground water. It was normal practice; if we didn't follow it, we'd lose our backing from the development committee and from the co-operatives. I no longer made decisions according to the needs of the soil or my animals. I was preoccupied with the possibility of managerial errors which could ruin us.'

Stuck in this spiral of production to please the bank manager, François and Françoise no longer cared what they produced, or how they produced it. Then, in 1980, brucellosis[2] hit their cows, and the epidemic destroyed the whole herd. They recognized this as an important sign that they should change tack and give up intensive farming. 'Françoise was expecting Benoît, our third child, and we wanted to spend time with the children. A farmer, like any other man, needs a family life. Going about things differently meant finding ways of being less dependent on the outside world, and thus introducing some coherence into what we cultivated.' With no help forthcoming from the Chamber of Agriculture, they approached other smallholders in the Manche who were abandoning factory farming, and soon came across André Pochon and his team at CEDAPA.

They had to go slowly, deintensify their farming progressively – it was like withdrawal for a drug addict – in order to keep the bank on their side. Their output was still dairy products, but they now worked on the principle of having no more animals than the soil could support; this also reduced the animal feed bills. The cows produced less milk, but the vet was called less often and the cows lived longer. Despite the savings, they could not make enough to pay for past mistakes or to support their family, to which Émilie was a further addition in 1985. They decided to produce veal; thirty-five calves at first, then fifty. Instead of falling back into the trap of factory farming, they set up a co-operative along with thirty other smallholders, which allowed them to keep control over the prices of the animal feed they bought and the cattle they sold.

FOUNDING THE FARMERS' CONFEDERATION

François Mitterrand's victory for the left in the 1981 presidential elections spurred Bernard Lambert to call for a regrouping of all farmers, with the aim of establishing a national farmers' union. François and the solidarity

---

2. Brucellosis is a serious infectious bovine disease causing abortion: it can be transmitted to humans in the form of 'Malta fever'.

committee responded to the call. There followed months of local debate around issues such as 'the discrimination inherent in European agricultural policy', 'the farmer's status', and 'the damaging effects of intensive farming'. François threw himself into these discussions, and came to the conclusion that agriculture was not reducible to mere production, but involved many interlinked aspects: 'Our job is not only to produce; we live on the land, we look after it and we are part of a rural social network.' Thus the 'sustainable farming' project was born, and the union established itself with strong bases throughout the regions.

The same year, François was elected Manche delegate to the CNSTP National Committee, where he met José Bové, a delegate from his region. This was the beginning of almost twenty years of comradeship, culminating in François taking charge of events after the McDonald's action, while José was in prison.

At about the same time, militants who had given up on the FNSEA set up another national structure, the FNSP. François Dufour and Bernard Lambert were sorry to see comrades, whom they would have preferred to be with them in the CNSTP, join this group. 'One day the two will merge,' declared Bernard. Unfortunately, a weak heart was to prevent him from seeing his prophecy fulfilled when the Farmers' Confederation was founded in 1987. At the conference at Bondy, 500 delegates, representing 70 regions, agreed on a new approach to agriculture to replace intensive farming: 'An agriculture that respects the soil and the environment, which is in harmony with the land, mindful of its crops and livestock, and creates decently paid jobs. An agriculture which takes the needs of citizens into account.' It was around this project that the CNSTP and FNSP fused to create the Farmers' Confederation.

In 1989, the Farmers' Confederation obtained 18 per cent of all votes in elections to the Chambers of Agriculture, giving them a significant representation. In 1995, they increased their vote to nearly 21 per cent, and François accepted the post of national spokesperson for the organization – a position where his knowledge, political instinct, convictions, curiosity, and reluctance to leave anyone by the wayside are well employed.

At home, he and Françoise had succeeded in creating the lifestyle they wanted, and the farm was working well. They applied to be classified as organic producers and, after the statutory three-year wait to conform to the soil standards, received the necessary certification.

PART TWO

THE DAMAGING EFFECTS OF INTENSIVE FARMING

4

THE BEGINNINGS OF
JUNK FOOD

There is no word in English that captures the full meaning of the French 'malbouffe'. 'Junk food' is the nearest English equivalent (and is used here), but in France, 'malbouffe' has overtones of unhealthy or contaminated food, and the sound of it provokes a feeling of nausea. This word was so apt that it caught the imagination of France, and its use spread like wildfire.

STANDARDIZING FOOD AND TASTE

'Malbouffe' – what do you mean by this word?

JB: The first time I used the word was on 12 August, in front of the McDonald's in Millau. While I was discussing my speech with friends, I initially used the word 'shit-food', but quickly changed it to 'malbouffe' to avoid giving offence. The word just clicked – perhaps because when you're dealing with food, quite apart from any health concerns, you're also dealing with taste and what we feed ourselves with. 'Malbouffe' implies eating any old thing, prepared in any old way. It hasn't been theorized, but it's become universally accepted to express a confused unease, a mixture of guilt and accusation.

For me, the term means both the standardization of food like McDonald's – the same taste from one end of the world to the other – and the choice of food associated with the use of hormones and GMOs, as well as the residues of pesticides and other things that can endanger health. So there's a cultural and a health aspect. Junk food also involves industrialized agriculture – that is to say, mass-produced food; not necessarily in the form of products sold by McDonald's, but mass-produced in the sense of industrialized pig-rearing, battery chickens, and the like. The concept of 'malbouffe' is challenging all agricultural and food-production processes. During the summer of 1999, problems with food went beyond health scares and entered the political arena.

FD: Today the word has been adopted to condemn those forms of agriculture whose development has been at the expense of taste, health, and the cultural and geographical identity of food. Junk food is the result of the intensive exploitation of the land to maximize yield and profit.

But surely at McDonald's, the food is clean and safe?

JB: At McDonald's, in theory, the hygiene regulations are very strict – sometimes more so than the current legislation requires. McDonald's has made this its trademark. The problem is one of dealing with frozen products. Their production is centralized: the hamburgers are made in Orléans, the chips are processed in Lille, the salads are assembled in Perpignan and the bread is baked in the Paris region. Everything is delivered by freezer-lorry to each McDonald's branch. Between the cold room and the sales counter, any breakdown in the cold chain could be dangerous. Branch managers have revealed that profits are paramount. So, for example, labels on the takeaway salads have been changed after the expiry date has passed. This was confirmed by an ex-McDonald's manager on Philippe Gildas's TV news programme.

There are problems with their drinks, which are sold not in tins or bottles but in cardboard cartons: ready-prepared juices directly plugged into the gas and water supplies. If there's a problem with the quality of the water. . . .

The hygiene standards are very strict, but the hunt for profits and speedy turnover means that these norms are sometimes bypassed, and things can go wrong. The food is completely uniform; the hamburgers have the same shape and content all over the world. In fact, it's 'food from nowhere', not even from a degeneration of American culture. Everywhere the same labels, the same way of running the 'restaurants' for a fast meal (though not necessarily fast service). It's the same with their immediate competitors, such as Quick or Burger King. That's why McDonald's represents anonymous globalization, with little relevance to real food.

Coming back to the McDonald's hamburger that you're so opposed to: it is, after all, made up of French meat, isn't it?

JB: This is the argument that always comes up. Despite its publicity campaign in autumn 1999, McDonald's uses French meat only because it's cheaper, not because of any consideration for the interests of French farmers. Before the 1996 embargo on English beef because of mad cow disease, McDonald's got their supplies from outside France, wherever it was cheapest. The meat they serve is made up of bits from the cheapest parts of the cow. All the burgers have the same make-up: fat is added so that the proportion of fat to lean meat is identical in every branch, ensuring that it can be cooked in the same way and have the same texture. Taste doesn't come into it: that's the reason for all the pickles and sauces. It's the same with their chicken meat, re-formed into 'nuggets'. In his excellent book *Petit manuel anti-McDo à l'usage des petits et des grands* (The Little Anti-McDonald's Handbook for Young and Old Alike), Paul Ariès describes this and many other things in compelling detail.

It's not just with fast foods that we see standardization. It's a general strategy of the food industry: to create the same tastes, so as not to frighten the consumer. The industrialized production of pasteurized cheeses is another good example.

Our society provides the consumer with processed food, ready to eat. It's increasingly rare for us to know the origins of our food: doesn't this encourage food scares?

JB: Of course it does. Considering the distance between the point of production and the point of consumption, and the increasingly complex processing and packaging procedures, the food product is rarely eaten in anything like the state it leaves the farm. It's reconstructed, often several times over, to produce easily prepared ready-made meals that can be consumed with little work in the home. The food industry regards the farmer as merely the supplier of raw commodities to meet the needs of the food manufacturers rather than those of the consumer. Ready-made foods and canteen-type restaurants are the two sides of profit's coin.

People may be concerned about buying ready-made meals, but they reassure themselves with the idea that they know what they're getting. It's both a time- and money-saver, and a cultural phenomenon. The art of cooking and eating together will soon not be passed on to new generations; this has resulted in a loss of family cohesion, and of the ties that bind us to the land or place where we live.

A number of groups – for example, one in the Ile-Saint-Denis, in the Paris suburbs – are researching these issues: the way we feed ourselves. By organizing theme evenings on taste and shopping for food, they aim to show the participants the importance of using natural produce. They invite people who live in an area, or tenants of a building, to make and exchange dishes native to their country or region. It's cheaper, better, and adds a *joie de vivre* to eating. Pleasure and food go hand in hand – or should do – but many families no longer eat together in the evenings.

According to dieticians and nutritionists, family meals occur less and less frequently. Everyone helps themselves in the kitchen according to their needs, grabs a frozen meal, bangs it in the microwave. Then they eat it in front of the television, often without saying a word to the rest of the household.

JB: It's a major loss. Apart from the odd Sunday, or special occasions, the meal is no longer the focus of the day, a time for conviviality and sharing. This change is due to the pressures of contemporary culture at work and during leisure time. It's also a symptom of the vacuity of much of modern life. Similar points can be made about birth and death. Death is no longer confronted, and the dying are no longer welcome at home but are sent to nursing homes or hospices. In fact, our deaths, like our food, have become standardized. It's the same with medical intervention at birth. Technology is stripping meaning from all of life's activities.

FD: Any general understanding of food and where it comes from is also rapidly disappearing. Nowadays, agriculture is increasingly devalued, and alongside this, food is increasingly standardized. We have completely distanced ourselves from an understanding of our relationship to food. People no longer see the link between wheat and bread. We eat, but we don't nourish ourselves, even on the farm.

JB: I've heard a story about a major cereal producer. He went bankrupt, and ended up visiting soup kitchens while still living on his farm. The very idea of a kitchen garden was alien to him. Unheard-of for a farmer! These single cereal growers have totally lost touch with the soil their livelihood was dependent on. Yet in the countryside, even under the most difficult circumstances, it's always possible to grow flowers and vegetables, rear chickens and fatten a pig. Their daily experience is no longer in concert with the reality of the land.

FD: It's been a while since many agricultural workers had their own vegetable patch and chicken runs. I remember a training course I attended in 1968, in the Loiret. It was at a farm with 130 hectares of cereals, but no chickens or rabbits. I asked them: 'How come you don't have some chickens for your own use?', and they replied: 'And what about our holiday month?' Worse still, I know farmers who daren't feed their children on the food they grow or rear. It's a worrying reaction from many in the industrialized chicken business.

In the space of fifty years, we've gone from the postwar shortages to today's food scares. Everyone agrees that we're eating more healthily, as our longer life expectancy proves. So are there really any risks associated with the degradation of food?

FD: The techniques and rhythms of production, the rotation of crops, have always been motivated by productivity. The example of animal feed given to herbivores is a case in point. For a long time the risks of this were known, and studies had indicated that all such feed should be cooked to eliminate pathogens. However, in order to be more competitive, these procedures were not respected, and now we see the results: an epidemic of BSE in Great Britain, and its transmission to humans. Health risks have been increased by a greater concentration of animals in small areas. So what do they do? They increase the use of antibiotics! We may think we're less exposed to the major microbial diseases than before, but I believe the risks are unknown and may be much more widespread, and it'll be some time before we know their full effects.

Food production is far too industrialized, and all sorts of ingredients – colourings, preservatives, stabilizers, water-retaining agents – are incorporated into its manufacture. As factories congregate close together, each carrying out different forms of processing, the consequences of an accident could spread like wildfire. This is why the 'traceability' of products is so important. It will allow action to be taken as quickly and as effectively as possible, whether it be listeria in the potted meat or cheese, dioxin in chicken, or traces of benzine in Perrier water. From the point of view of dealing with potential dangers, this may be effective, but the consumers' fears have not been allayed. The risks remain, and the number of food scares will increase. As long as profit and productivity are the main motivation, careful and natural forms of working will disappear.

This industrialization of food production . . .?

JB: For the great majority of producers, it occurs when the price of the ingredient they produce is being negotiated. Let's look, for example, at how the retail price of cow's milk is arrived at: the criteria have nothing to do with the quality of the milk. The cost of production is hardly reflected in the final price. The industry ensures that there are Chinese walls between production and consumption.

The specialization that accompanies agricultural industrialization means that the global dimension is lost from sight. This applies equally to the farmer, the consumer, and all the other workers in the food chain. The standardization of mass production, the division and splitting up of the work, are the consequences of unacceptable shifts and changes which, we believe, are linked to the type of modernization imposed on agriculture after the Second World War: it's what we call productivism, or intensive farming.

FROM SELF-SUFFICIENCY TO AGRO-EXPORTS

You're very critical of the way French agriculture has developed; yet when we see the progress achieved since 1945, you might think our agriculture has nothing to be ashamed of. With the exception of a few more tractors and the arrival of electricity, the French countryside at the end of World War II was little different from what it had been at the beginning of the century. So farms were small; many had no proper water supply, and little mechanization. They were so unproductive that the government was obliged to introduce rationing from January 1946 to February 1949. It was not until 1950 that the volume of production equalled prewar levels, and this was far from making France self-sufficient in food. Now, over fifty years on, not only are we self-sufficient but we are ranked second to the USA, as an exporter of food.

FD: It's a real success story, a European success. On the strength of the reconstruction of Western Europe, the establishment of the Common Market in 1957 and agricultural legislation in 1960 and 1962, the modernization of agriculture has leapt ahead. The objectives were clearly outlined: to achieve food autonomy in Europe, to provide foodstuffs at the lowest cost, and to protect a number of European agricultural products – cereals, sugar, milk, meat and wine – from international competition. Another consideration was attempting to guarantee farmers the same income as townspeople. To attain these objectives, France chose to move towards industrialized modernization: specialization, concentration of farm ownership, and the creation of complex chains of production.

But though these policies have met with considerable success, we need to see what the cost of all this is. Although agricultural activity still accounts for half of all the land use, today only one in ten people earn a living from the land compared to fifty years ago. Much more is grown and reared per hectare than before, but in most instances we've turned our backs on traditional farming's respect for biological rhythms. Farm specialization has also created monocrop regions, causing demographic, economic and ecological disequilibrium. Cereals have replaced livestock in the big fields; breeding is now concentrated in the west of the country, and has been sustained only with difficulty in mountainous areas. Consolidation of farms, draining of the land, cutting down hedges and levelling of hills have all taken place in the name of economic gain, without any heed – at least, not until very recently – to geographical, hydrographic and climatic considerations.

JB: The rationale is that all agricultural policy since 1957 has been geared to low food prices and food self-sufficiency. Food security for Europe was an essential and legitimate political goal. The problem is that once self-sufficiency was achieved, the policies didn't change. We could have maintained our self-sufficiency at a European level without getting embroiled in this excessive industrialization, whose only purpose was to produce for the sake of producing, with the EC responsible for finding outlets for the overproduction, or for compensating the producers.

You describe the tasks assigned to agriculture and its farmers, but these in themselves don't explain the particular form of agriculture adopted and the fact that, according to you, it is an industrialized or intensive one.

JB: The arrival of the first soya beans in French ports, not subject to any Customs duties, signalled the start of agricultural industrialization. When the Common Market was established in 1957, the six member states allowed this import as a major concession to the United States, to whom we owed the aid of the Marshall Plan and the defence umbrella in the face of the Soviet bloc.

Cheap soya beans are very useful in intensive breeding, because they make it possible to rear herds on small areas of land close to the delivery ports in the Netherlands, Belgium, Denmark and Brittany. Later on, the soya beans came from Brazil.

There is now an absurd cereals policy that assigns a ridiculous prominence to maize grains and silage.[1] Cereals[2] are very rich in energy, but poor in nitrogen; hence the need to provide a food supplement for animals fed on cereals in the form of soya beans imported into France by American multinationals. So now the major market for European cereals is no longer

1. When the whole plant is harvested, maize becomes fodder. Silage is a means of conserving the fodder in its green state, by fermentation in airtight silos.
2. The most commonly cultivated cereals are wheat, barley, oats, corn, wheat durum (used to make pasta). Grain maize is classified as a cereal.

human food – more than 75 per cent of cereal production is destined for animal feed. This is why, today, there's a shortage of wheat production for bread-baking. This consideration was completely ignored by the cereal lobby.

When the Common Market was created in 1957, France saw herself as the granary of Europe, and it was from this standpoint that she negotiated for a Common Agricultural Policy (CAP). The agricultural lobby was already very powerful, and cereal growers obtained a very good remuneration guarantee from the Common Market. They also got rid of the 'quantum' (a system of guaranteed prices, graduated according to the amount delivered by the farmer) which had been established in 1936 by the Popular Front government. The quantum system had helped many farmers out of the dire crisis of the 1930s. The CAP guaranteed cereal growers high prices within the Common Market, and protected them from the world market by a system of adjustable tariffs on competing imports. Above all, their income from exports was guaranteed at internal prices, any shortfall of the world price below the European being compensated with subsidies. The cereal and beetroot growers benefited most from the CAP, which was conceived as an answer to the postwar deficit in food production.

This policy, however, very soon led to an overproduction of cereals, which had to be sold on the international market. At the same time, we continued to import – tariff-free – soya beans and cereal substitutes, such as manioc from Thailand, as these provided cheaper fodder for livestock. The subsidies given to exports – financed by European taxpayers – were totally perverse, as they fuelled speculation on the world commodity exchanges and became a barrier to attempts by developing countries to be self-sufficient in food. So this agricultural sector is based on artificial markets, and sells at artificial prices.

FD: It would be interesting to analyse the global cost of this type of agriculture. Our predecessors were told to keep an eye on the household shopping basket, to focus on producing cheap food; but now, what the consumer saves on buying meat from stock reared on imported soya beans, she or he pays back in tax to subsidize the cereals. Add to this the cost of destroying existing crops to plant cheap soya beans in countries such as Brazil, together with the social, economic and environmental damage caused by this industrialized agriculture. None of this has previously been included in the cost of basic foodstuffs. You can certainly buy chicken at under ten francs a kilo, but do you know the real price paid by society?

AN INSIDIOUS REVOLUTION

This agricultural modernization can't have resulted solely from CAP measures. Technical and structural changes were needed to obtain the required output. How was this achieved?

JB: I believe the fundamental idea which underpinned the modernization of agriculture in the 1950s and 1960s, is the same that applied in industry: intensification and specialization of output, rationalization and segmentation of work, standardization of product. With its scientific organization of work, industry became the reference point for measuring economic efficiency. But it was felt that in agriculture, in contrast to industry and commerce, these transformations could be assimilated to the social objective of maintaining the family farm.

In fact, this objective was difficult to reconcile with the overall project and its reliance on specialization, the key concept in modernization. Specialization of land use from areas for breeding livestock to areas for large-scale crop growing, from cultivated fields to woodland areas. Not only is there specialization of individual farms, but, in the space of a few years, we've moved to monoculture in entire regions. For example, in Brittany, until the 1780s, a farm would produce milk and either pork or poultry; today it has a single specialized output. Dairy farmers are no longer all-round farmers;

they're specialists in milk production, with little interest in crops and even less in soil use. This specialization has put an end to local production of different crops and animals, which are adapted to the climate, soil and topography of the area.

At the same time, agriculture has adopted a production-line organization involving segmentation. In poultry rearing, for example, one farm specialized in the production of one-day-old chicks, another in larger chicks, and yet another in laying hens. With bovine production, you'll find suckling cows in the Massif Central and workshops for fattening up veal calves in Brittany, in Champagne-Ardenne and in Italy. Between each link in this agricultural chain, an intermediary buys or sells the products. It may not be the conveyor belt, but it has a lot in common with it.

JB: Mechanization has played a crucial role in the intensification of agriculture. From the beginning of the 1950s, the leaders of the JAC saw it as the future for the countryside, and offered tractors and other machinery to enable agricultural workers to gain parity of income with the rest of society. Today, it's not uncommon for a farm to have several tractors. It's paradoxical to see agriculture, which is reliant on the natural use of solar energy, increasing the consumption of non-renewable fossil fuels. If you add to this the energy consumed in producing food and transporting it across the world, the balance sheet is even more negative.

People talk about the 'silent revolution'. Why silent?

FD: The farmer who continued to use his own methods was made to feel guilty by the revolution in agricultural techniques. Knowledge came from outside, and totally devalued the farmer's know-how. In the name of freedom and emancipation, he had to make a clean break with his former practices. Instead of being a farmer, he became a 'producer', scrupulously applying new techniques under the guidance and control of technicians. At the end of the 1970s, when I was doing my agriculture training, all we had drummed into us was a technique known as 'maize–soya–concrete'. The fields were to be ploughed up for the maize crop; soya was to be bought to complement the feed for animals that were to spend the whole year in a modern concrete barn rather than led out to graze in the fields. The lecturers had no qualms about denigrating the methods used by our parents, such as crop rotation or the use of varied fodder and permanent meadows for grazing.

JB: In this way, agriculture became a vast market for all sorts of subsidiary industries: manufacturers of building materials and equipment, energy, fertilizers, animal feed, veterinary produce – not forgetting the work of technical advisers. And the Crédit Agricole bank was given the role of providing the loans to gear up the modernization of agriculture, and to pilot the work and its techniques.

So we're dealing with an agriculture chasing after techniques rather than techniques being in the service of the farmer? Have the fields been adapted to the machines?

JB: Absolutely! The size of a plot of land has been adapted to the machine, often to the detriment of the natural topography and the needs of proper drainage. Hedges that hindered the movement of machines and competed with crops were uprooted, and slopes were flattened. All this reshaping by bulldozer has resulted in a loss of biomass, promoted soil erosion, reduced the humus layer, and significantly decreased the flora and fauna, as hunters will testify.

Today, there's a return to putting animals to pasture, but there's no shelter from wind or sun, and planting new hedges to make windbreaks is costly. Recently there has been a move away from some of the more extreme practices of intensive farming. Modifying the ecosystem by draining has caused considerable damage to the wetlands, as evidenced by the swamps in Poitou. Over the last few years, the European Union has tried to protect these zones through the 'Natura 2000' directive. Its implementation in France has met with opposition from local landowners and hunters. All over France, draining has been the means of reclaiming land for cultivation, to the detriment of wildlife.

AN OBSESSION WITH MAIZE

You give a prominent place to maize in this industrialization of agriculture. Isn't it a bit obsessive?

FD: This plant symbolizes intensive farming: it's a favourite crop, because it's highly successful, but it causes a lot of damage. It's a summer plant with considerable productive potential: if the soil is well fertilized, yields easily reach 80–120 quintaux[3] per hectare, whereas wheat yields only 60–100 quintaux on good soil. So, under the same conditions, more maize can be produced and more animals fed. Maize can be more heavily fertilized, notably with nitrogen. This explains why it's called a 'dustbin plant': growing it can pollute water supplies and the environment.

JB: American agribusiness was very quick to catch on to the fact that substantial profits could be made from maize – the plant the indigenous Americans gave to the first settlers. In the 1920s and 1930s, the government introduced a hybrid maize which forced the farmer to buy new seeds every year. In France, the public research institute and the seed companies poured a lot of money into maize. When it first crossed the Atlantic, the optimum region for this crop was the southwest of France. But as it was improved genetically, it made dramatic strides into the north of Europe. Distant relatives of the plant are today found in Great Britain, in Germany and in Belgium, thanks to the process of hybridization and the chemicals it needs being dumped over the fields.

3. One quintal equals 100 kilos.

The environmentalists denounce maize as a big consumer of herbicides and pesticides. Is it the worst crop for the environment?

FD: It all depends on the place it occupies in the rotation of crops. Its attributes favour its use in monoculture, the highest stage of specialization in crop growing. The spread of maize monoculture, which is very demanding of 'inputs' of fertilizers, herbicides, and pesticides, has adversely affected every aspect of the environment. The soil, plants and parasitic insects have developed resistances to these chemical treatments. In winter you often need to weed the fields completely to prevent an invasion of parasite-resistant plants.

Maize, being a summer plant, needs a lot of water. In some areas planted with this crop, agricultural irrigation uses 80 per cent of the water supply. Moreover, this irrigation is largely subsidized, with public money available for drilling, reservoir building, and the installation of pumping systems. The farmer pays barely 10 per cent of these costs.

In the livestock-breeding areas, the intensive approach consists of planting maize in the summer so that the ground is not left barren, and to obtain maximum yields. In winter, fast-growing ray grass is sown; this also needs a lot of nitrogen. So there's a harvest in October and one in April, creating more work for the farmer. Soil that is producing two crops which are greedy for water loses its humus, becomes heavier, requires stronger machinery to till, and is more susceptible to extremes of weather. Eventually the ground just gets clogged up and fulfils no function other than holding up the plants: the rest is done by chemicals.

JB: In intensive farming the object is to adapt the soil to the crop, never the other way round. This is a fundamental change in agriculture.

Specialization has also been applied to livestock. Rustic breeds are disappearing, giving way to genetically selected breeds. Are there risks in this?

FD: The most noticeable phenomenon one sees while travelling through the countryside is the predominance of certain breeds, such as the Holstein black for milk or the Charolais for suckling. Genetic improvements, in animal as well as crop production, have led to a disappearance of local breeds that were adapted to the microclimate, soil and fodder of the area, but had the disadvantage of not being sufficiently productive. There are essentially two objectives in this selection process: an increased yield, regardless of quality; and ease of breeding, so that the farmer can have a bigger herd. In milk production, for example, not only has the yield of each animal been increased, but thanks to mechanization and automation, so has the number of cows.

If you drive your car at breakneck speed, with savage acceleration, it won't last so long. The same is true of intensively reared cattle. Farmers are encouraged to drive their herds to the maximum of their capabilities rather than to treat them prudently, with due regard for their health. Intensive milk production means that a cow, on average, has about three calves in a lifetime of a little over five years. With our form of sustainable farming, cows live for more than ten years.

JB: It's the same with sheep production. Intensive farming requires a renewal of 35 per cent of the sheep stock annually. The sheep farmer who doesn't do this is not up to his task, and is considered unprofessional. Consequently the system is pushed to the limit, and animals are made to reproduce from their first year, then sent to the abattoir, on average, at the age of four.

FROM JAC ENTHUSIASM TO TODAY'S
'JUNK LIVING'

Traditionally, the farmer's livelihood is based on his relationship with living things; a farm is in harmony with nature — this demands a real knowledge of the way plants and animals grow: a sensitive approach to the rhythms and cycles of nature. So how do today's farmers deal with the logistics of industrialization?

JB: Once again we need to look back to the late 1950s and early 1960s, and recall the crucial role played by the JAC in this development of farming in France. Set up just before the Second World War, the JAC rooted itself in strong Catholic areas, where it became a means of emancipation for young farmers. Their ambition was for parity with labourers — parity as much in terms of respect, and a place in society on a par with the rest of their generation, as of money. The JAC told them: 'You're not bumpkins, you're young people, like the others. Be proud to be farmers.' This went hand in hand with a desire for modernization, so that farmers didn't look like a backward sector of society. The JAC formed a whole elite of youth, keen on the ideal of reconstructing society in the aftermath of war and conscious of being leaders in their field. These young people, our elders, totally immersed themselves in the logic of agricultural modernization, convinced that it would result in the emancipation of their communities. The establishment of the Common Market in Europe, and the changes in agricultural laws in France, enabled them to obtain modern tools, to participate in intensive farming and, simultaneously, to become independent of the family. Borrowing started — first to buy a tractor, then to build a house after leaving the family farm where two or three generations had lived. Young people soon found themselves in a spiral of debt. Farmers were turned into consumers, like everyone else. Formica invaded farm kitchens just as it invaded urban workers' homes.

FD: The JAC modernization aimed to overthrow the old order of farming culture — the world of a reactionary and moralizing Church, of landowners, animal merchants and other petty speculators. It allied with the God of Progress in a three-stage programme: technical innovation leading to economic growth which, in turn, would create social progress.

Were all the young farmers caught up in this turn towards intensive farming?

JB: This economic model didn't develop everywhere at once. Only farmers in certain areas where the land conformed to the needs of industrial farming were able to get credit, and technical aid. There were whole swathes of the country where the majority of farmers stayed outside this development: all the mountain regions, the areas where farming was difficult. Intensive farming took hold there only later.

But hasn't modernization improved the farmer's life by reducing the length and arduousness of the job, and by increasing his income?

FD: Well, that was the original idea, but once you get deeply involved – once you've invested in a specialization, once you go into restructuring schemes to pay for an industrial building or some very costly piece of machinery – the farmer has to produce more. In order to do so, he has to take over his neighbour's land and production. The word 'farmer' became a misnomer; the farm's functions had been diverted from their original purpose.

JB: The farmer's job is a combination of manual work, reflection and creativity. Today – with off-soil workshops, integrated rearing, specialization of production – the farmer has become a slave to the system, and the job has been vastly impoverished.

But don't today's farmers enjoy the same standard of living as people in towns?

JB: To promote sales of tractors, dealers constantly reminded farmers of how hard their job was – that they had to get up very early, around four in the morning, to care for their animals, and then had to go off to the fields. They tried to make farmers believe their work would be less arduous with mechanization. In reality, farmers don't work any less, just differently, and they're much more stressed.

Two films dealing with this issue – *Farrebique* and *Biquefarre*, directed by Georges Rouquier – are very revealing. The first film describes a farmer's life after the war; the second is set in the 1970s and 1980s. The change couldn't be starker. In the first film, you can hear birds, nature, people on the farm. In the second there's merely noise, and you can see the illness resulting from stress linked to intensive work – cardiac problems and hypertension. Farmers' work is as hard as ever, but they're no longer in touch with their roots. The slightest mechanical breakdown throws them into complete disarray. The balance sheet is that although living conditions have improved, and there is more comfortable housing, family life has deteriorated. One of the most telling signs of this is the shortage of new generations of farmers. Young people are just not interested in living the life they've seen and experienced on their parents' farm.

FD: Family failure and social isolation have become commonplace. Among farmers who have settled in the last twenty years – including those in the west of the country where, in theory, modernization was supposed to maintain farmers in thriving communities – a significant number remain unmarried.

JB: Some years ago, single farmers were those who had stayed on the land because they couldn't do anything else. Today, the majority of farmers who are thirty-five to forty-five years old and single are those who've had the most training. They are the young people who entered the race for production and who woke up one day to realize that they were alone. And, alone at forty, they have breakdowns, give up.

THE FIRST BREAK

When did farmers first begin to realize that their income was not what they had hoped and that there wasn't enough room for everyone?

JB: Towards the end of the 1960s, the Western branch of the FNSEA and its main leader, Bernard Lambert, pointed out the discrepancies between the fortunes of farmers in regions with mainly small-scale production, a long way from large consumer centres, and those with large-scale cereal farms in the Paris basin, who benefit from very high prices and have no restrictions on their output. This disparity was at the root of the establishment of the Farmers' Confederation.

From the start, the modernization of agriculture, so lauded by the FNSEA and CNJA, depended on a massive exodus from the countryside, and for those who had dared to invest, the hoped-for profits were not forthcoming because of mounting debts.

FD: Thanks to off-soil animal breeding, farmers in poorer areas were able to stay on small farms, and if the bank was unwilling to lend, a feed manufacturer would often offer an 'integration contract', with the firm supplying the animals and their feed. That's part of the reason why the issue of off-soil animal breeding is so important. Even today, on average-sized farms, the construction of a battery chicken shed provides sufficient income for the young farmers to leave home. Today it's obvious to us that alternatives to battery farming are preferable, but this was not the attitude of young farmers seeking to escape grinding poverty in the 1960s. They formed dissident groups in the farmers' organizations to fight against the bigwigs, for the rights of the small farmer.

JB: But even those who, like Bernard Lambert, broke with the prevailing attitude in the FNSEA didn't challenge the fundamental issue: intensive farming. They wanted a better distribution of its benefits, they fought against the drop in the number of farms and the rural exodus, but they didn't question the basic premises. Not until 1980 did Lambert fight against hormone-fed veal, and in 1982 the CNSTP made an official break with such farming methods.

A SINGLE UNION AND
JOINT MANAGEMENT

You've frequently mentioned dissent within the FNSEA, and the role played by this union in the 'modernization' of French agriculture. What has that role been precisely?

FD: The transformation of French farming can't be explained simply in terms of the enthusiasm of the JAC, which in any case shrank after May '68. In less than forty years – barely two generations – the number of farmers in France has declined fivefold. The full explanation for this can be found only if we take into account the establishment of a single union.

JB: The FNSEA was set up in 1946, from the ruins of the Farmers' Corporation, which had been imposed by the Vichy regime. A vertically structured system, with membership practically mandatory, this organization took charge of everything: purchase of produce (or supplies), distribution of ration vouchers, and so on. In the name of 'Farmers' Unity' – the title of a famous speech by Eugène Forget, its first president – the FNSEA was the only union for farmers. All farmers, regardless of differences in status (landowners, farmers, tenant farmers, breeders), and regardless of the size of their holding and what it produced, were supposed to have the same interests, and were therefore best represented by a single organization. From this position, the FNSEA managed agricultural policy jointly with the government.

What did this amount to? Even at that time, there were numerous references to 'the end for farmers', or 'France without farmers' – the FNSEA leadership used the political weight of this single union to negotiate agricultural policy in secret with ministers, and then control its implementation on the ground.

FD: Farmers were under the control of the FNSEA or its appendages in every sphere of their professional life: obtaining financial help when they settled in, arranging certificates under the milk quota scheme and loans from Crédit Agricole, as well as buying insurance. The farming press, whose advertising revenue depended on intensive farming methods, also supported the union.

The single-union regime gradually developed into an 'extended family' grouping – not just the farming organizations but also the insurers, the bank (Crédit Agricole) and the co-operatives. This 'family' agreed on one specific point: the pursuit of intensive farming, on the back of which these organizations were founded.

It is worth noting that today, Crédit Agricole is the number one European bank and number three in the world, on the basis of the funds it holds. It still finances 80 per cent of farmers in France. One in three French people bank with Crédit Agricole. Clearly, it's unlikely that the policies of such an organization will be in the interests of ordinary farmers. But it's not always easy to avoid the pressure of mandatory membership, or having the affiliation fee deducted from one's current account by the co-operative or other affiliates.

Since 1981, pluralism in union representation has been officially recognized, and representatives of minority unions have been imposed bit by bit on the 'extended family'. The direct consequence has been a gradual loss of influence for the FNSEA and CNJA. In the last elections to the Chambers of Agriculture, the FNSEA and CNJA polled barely 60 per cent of the farmers' vote, whereas the Farmers' Confederation had just over 20 per cent. To press for change in the politics of agriculture is to challenge the whole system of joint management.

At what point did people start talking about France's need to export rather than food self-sufficiency?

FD: By the late 1970s, France and Europe had achieved the objective of self-sufficiency set twenty years before, at the beginning of the Common Market. In fact, it was not long before Europe was weighed down by mountains of surplus butter, powdered milk, cereals and beef, all heavily subsidized by the CAP. Europe is now financing the cheap sale of these products on the world market, with export contracts to Russia and the Arab countries, for example.

In 1978, in Vassy (Calvados), President Giscard d'Estaing emphasized France's commitment to food exports by referring to agriculture, at the height of the oil crisis, as France's 'green petrol'. In reality, this strategy is reduced to producing for the sake of producing. You can't put the brakes on a machine that's so profitable for the whole of agribusiness. Certainly, there's little resistance from agriculture's joint management, which ignores the challenge to intensive farming from farmers and others, and is blind to the explosion of the European budget – financed by consumers paying VAT. France's only concern is for a fair return on its contribution to the European budget, and French agriculture is dramatically impoverished as a result.

CO-OPERATIVES IN CRISIS

So French agribusiness went out to conquer world markets, and encouraged its subsidiaries to produce more and more. The co-operatives fell in behind. They appear to have moved from being organized for the collective interest of farmers and consumers to the interest of agribusiness and its managers?

JB: In accordance with the principle 'Unity Creates Strength', farmers have been involved in setting up co-operatives for the purposes of buying, selling or providing mutual services since the end of the nineteenth century. While the postwar boom benefited the co-operatives, it also created competition between them; some grew, often by buying up other companies or co-operatives. The farmers remained at the head of these growing businesses, but were unable to look critically at what they were doing, and were often completely mesmerized by the trappings. They went along with the system, and were then overtaken by it. Farmers and community figures are represented on administrative councils, sometimes even taking the presidency, but in general they have little management authority. The managing director runs the business with little consideration for the collective nature of the company, or for solidarity between those involved. This sort of co-operative then develops a dynamic of growth where the farmers and members are seen only in an economic context. They are both a burden, due to the price of their produce, and a commercial outlet for all sorts of goods and services that the co-operatives can sell them.

FD: Along with the modernization of farms, which was supposed to bring greater equality of income to farmers, the FNSEA also presented the co-operatives and Crédit Agricole as tools for winning economic power, with investment at the onset of production and the commercialization of agricultural goods. In reality, the co-operatives were forced to behave like private industries, working strictly within the constraints of the market.

We can see the most obvious failure of this quest for economic power when we look at the way large-scale distribution can use its economic muscle to dictate prices and terms to the agricultural sector. The economic relationship of forces is fundamentally unfavourable to the food industries, and will remain so for the foreseeable future. Farmers can avoid selling their produce at a loss only if the state guarantees minimum prices. In the absence of such measures, which would involve an entirely different agricultural policy, we need more transparency in the distribution organizations' transactions, and farmers should have the right not to sell produce at a loss.

JB: The whole spirit of co-operatives changed when they were reorganized in the pursuit of world markets. They became part of an industrial machine, and were told: 'These are the rules of the game; you can't change them.' Agribusiness imposed increasingly competitive demands on the co-operatives, employing performance-related criteria. These schemes are concerned more with the constraints of manufacture than with the quality of the product. They are also a way of determining which farmers will be allowed to remain in the co-operative. This is an ongoing battle. Co-operatives such as AOC Reblochon and Beaufort, which produce cheeses, or the Jura fruit co-operatives, have blocked the big dairy groups. Human-scale co-operatives such as these allow for both direct democracy and the practice of an agriculture linked to one product and one area.

5

FARMING AGAINST NATURE

A basic tenet of traditional farming held that the farmer knew how many animals he could put out to graze on a field without damaging its natural cycle of regeneration. So he would know that he could graze two cows per hectare, or five sheep, but only a single horse. The number of hens was limited to the amount of grain the poultry yard could accommodate, and pigs were fed on potatoes, barley and milk. This agriculture was in harmony with nature, and the men and women in its service.

Modern agriculture, however, has turned against nature, and left this type of farmer behind. He has had to make way for the engineer, the technician and the builder. Instead of using the natural rhythms of the land, as they do in the East, this intensive farming destroys them in pursuit of its own priorities. It has built itself a technical and chemical arsenal, the damaging effects of which we are only beginning to discover. The only brake on this agriculture freed from natural constraints is the pressure from consumers who find on their plate beef fed on lamb remains and the residue of septic tanks, or vegetables modified by animal or human genes.

DOPING DOWN ON THE FARM

From the Worker-Farmers to the Farmers' Confederation, your movement has fought against the use of hormones in veal, pig meat or poultry; you're opposed to them all. Why this total opposition?

FD: The sole purpose of using growth hormones, natural or synthetic, in raising livestock is to maximize the yield from the animals. This is done regardless of the consequences for the health of the consumer, or the quality of the meat. The use of hormones is just another way of making food production artificial, against the natural rhythms and cycles of animal life. It's one of many techniques that intensive farming has adopted and our union has always opposed.

In 1980, the Worker-Farmers campaigned against the pressure being put on veal farmers, who were threatened with bankruptcy if they didn't use banned growth hormones. This led to a boycott movement launched by the consumer organization UFC – What choice? and forced successive Ministers of Agriculture to impose strict rulings on the use of 'growth activators' in farming, despite pressure from the European – and especially American – pharmaceutical industries, which envisaged lucrative profits from this trade. In 1988, farmers and consumers won a victory when the European Union prohibited the use of all hormones in livestock rearing. This led to the trade conflict with the USA that we've talked about already.

JB: The example of bovine somatotrophine (BST) is relevant here. Monsanto, the world pharmaceutical giant, invested considerable sums of money to develop a hormone that could increase a cow's milk yield by up to 25 per cent without the need for any additional feed.

American and Canadian farmers have been allowed to use this hormone for the past few years. At the end of the 1980s, the EU was besieged with requests to allow it on to the market. Advertising in the agricultural magazines enthused about the benefits of BST, emphasizing particularly the positive impact it would have on farmers' income. In the face of this onslaught we alerted public opinion by various means, including the occupation of the headquarters of Monsanto-Europe.

In 1994, the EU recognized our social and economic arguments, and postponed putting BST on the market. It was clear that the use of this hormone added nothing to the quality or taste of the milk; on the contrary, there were risks of nutrient deficiencies.

Given that Europe controls the milk yield by using quotas to avoid overproduction, the only result of introducing the hormone would be to get the same amount of milk from fewer cows and farms, while at the same time, providing more business for the pharmaceutical industry.

FD: Today, the pharmaceutical companies, hoping to get the EU to revoke its 1994 decision, have changed their tune. BST is no longer a hormone – because the word itself provokes fear – but a natural, efficient and inoffensive protein. Trapped between WTO rules and the 'Codex Alimentarius',[1] and the interests of consumers and farmers, the EU has no clear position on hormones.

These rules of international trade provide for restrictions on the free trade in agricultural and food products only if scientific research has proved that they are dangerous to human health. The free market pays little heed to the need for caution, still less to social and environmental considerations. We're dealing with a market, aided and abetted by science, which works against people and the land.

Whatever the misgivings of the EU about the health and social risks in the use of these products – hormones, GMOs, antibiotics, and so on – it and its member states are more fearful of massive consumer food scares and the consequent effects on their economies.

---

1. The 'Codex Alimentarius', established in 1962 by the FAO (the UN Food and Agriculture Organization) and the World Health Organization, is a set of directives designed to protect the health of consumers and guarantee good practice in the food production industry.

Are hormones still used today?

FD: In theory, they're forbidden, but all over France the courts are regularly sentencing farmers whose livestock has been found, at the abattoir, to contain hormones. The Farmers' Confederation has regularly sued as a victim of the crime, in order to follow the proceedings of the court, but we've rarely succeeded in uncovering where the subsidiary production and distribution of these banned chemicals takes place. The dealers are seldom arrested or sentenced. We suspect that these networks are assisted by the turning of a blind eye at official levels, including those charged with policing the ban. What we know about is only the tip of the iceberg.

Are these hormones dangerous to human health?

FD: We've known for a long time that certain synthetic hormones, notably those containing chemicals not naturally present in animals, can be carcinogenic. These hormones are banned virtually everywhere in the world, but that doesn't mean there's no traffic in them. Like drugs in sport, some are very difficult to detect. Other hormones are dangerous only above a certain dose. Many of those at present deemed inoffensive are also far less efficient, and this encourages heavy use. Again, as in competitive sport, the use of these drugs in animal rearing is dictated by short-term economic gain. Dangerous substances will continue to be used unbeknown to the consumer. Scientific experts in the EU have shown that some American beef sold on the world market contained hormones that were prohibited in the USA itself. By calling for a total ban on these products, we're reclaiming the principle of respect for the natural cycle and rhythm of agricultural production.

JB: As in the campaign against GMOs, I firmly believe that we won't be able to muster up an effective opposition to the pharmaceutical giants on our own. Commercial lobbies of all sorts exist, and have considerable means at their disposal to apply pressure on political decision-makers of every persuasion, even pitting them against each other. We can resist only by joining forces. That's why, in 1992, we set up the Farmers', Ecologists' and Consumers' Alliance.

In January 1996, the USA submitted a complaint to the WTO about Europe's refusal to import hormone-fed meat. In the background, Canada, Argentina, Australia, New Zealand and South Africa also had an eye on the European market. They were encouraged by the outcome of an international scientific conference organized by the European Commissioner, Franz Fischler, which had concluded that five types of hormones used in animals were innocuous. On 12 January 1996, as part of a European campaign for hormone-free meat, we brought Gertrude and Laurette, a cow and her calf, to the Museum of Natural History in Paris. The point was to show how these two animals would become living relics if agricultural production continued to be industrialized. As well as the animal breeders of the Farmers' Confederation, the demonstration rallied activists from the Farmers', Ecologists' and Consumers' Alliance, and prominent scientists who denounced the fraudulent practice of artificially bulking up meat with hormones: 'Water sold at meat prices'.

The following year, the European Parliament renewed its ban on the import of hormone-fed meat. Our campaign had forced the FNSEA to toe the line, even though for a number of years, in the name of progress, they had been in favour of the use of hormones. The co-operative bosses were impressed by the fact that using hormones could increase an animal's weight by 5 to 10 per cent, sending a cow's value up by 500–1,000 francs.

The issue of allowing the use of hormones in meat production comes up again and again, because there are powerful financial interests at work: partly those of the stock-breeders but, more importantly, those of the pharmaceutical industry. Later participants in the chain also have an interest, from abattoirs to the frozen-food business. An increase in profit margins for these companies is achieved through the standardization of carcases and the automation of butchery, all of which is aided by hormones and other anabolic drugs.

Are there any hormone-free growth promoters?

JB: Indeed there are. Antibiotics are essential medicines for human and animal health, but in some husbandry regimes they're used for other purposes: as growth promoters, to resist microbial infections, to compensate for the poor general health of the herd, and because an animal's own immune system has been weakened. In addition, there's the intensification of farm productivity: a cow 'pushed' to produce 8,000 to 10,000 kilos of milk a year becomes much weaker, and more susceptible to all sorts of health hazards, than a cow producing 6,000. The problem is more serious still when it comes to off-soil farming. In pig production, for example, it has been noted that vet fees per sow were proportionally much higher according to the size of the herd; they're more than twice as much for herds of 300 sows compared with those of 60 to 80 animals.

Most of the time these pharmaceutical food supplements are administered without vets controlling the choice or amount of antibiotics. Thirty years ago, resistance to antibiotics in humans was unheard-of. Today, it's an increasing worry. This is due to overuse in medicine, but also to their abuse in farming. A few years ago in Denmark, for example, a sick woman needed surgery. During treatment, the doctors discovered that her illness was resistant to essential antibiotics, and she died as a result.

FD: The introduction of European directives for the withdrawal of certain antibiotics from animal feed threw the spotlight on to the problem of the size of breeding farms and the methods they used. Rabbit farming, in a very short space of time, has been industrialized along the lines of poultry farming. Goodbye to the bucolic pictures of rabbits nibbling grass and bounding around the fields. Now there's battery breeding of hundreds of rabbits in a confined space, fed on meal and processed granules. A 'genetic improvement' has been introduced which, in under ten years, has reduced the fattening-up time for a young rabbit by more than 20 per cent – a real triumph for modern technology!

Hundreds of farmers who believed that large-scale rabbit farming offered a way out of the economic crisis they faced ended up deeper in debt, thanks to Crédit Agricole and advice from the other interested parties – feed firms, abattoirs. An outbreak of the devastating intestinal disease enterocolitis destroyed up to 70 per cent of the stock on some farms. The researchers and technicians who were so eager to organize the intensification of rabbit farming have been unable, over the last two years, either to diagnose the cause of this intestinal disorder or to get rid of it. The short-term solution has been to increase the dosage of antibiotics. Some farms have resorted to antibiotic treatments strictly reserved for humans. It's a very serious situation.

As a result of all this industrialization, rabbit farming has reached an impasse. The breeders affected should not have to suffer all the consequences themselves, nor should they put the health of consumers at risk in a struggle to survive. Collective responsibilities should be worked out in order to provide a reconstruction plan for these breeders: to compensate for their losses, for which they're not responsible, to farm rabbits differently, or to find an alternative product.

GMOS: CLAIMING A ROYALTY ON LIFE

On 7 January 1998, you destroyed a stock of GM maize owned by Noveartis in Nérac. On 5 June 1999, you repeated the offence against GM rice, growing in Montpellier. In November 1999, from Washington to Seattle, you both joined American farmers and consumers in demonstrations against 'Frankenstein food' – food containing genetically modified organisms. Why are you so hostile to GMOs?

JB: We're opposed to genetic modification in agriculture. It results in plants whose natural developmental 'programme' has been changed by intervening in its genome. The genome is all the genetic material characteristic of a particular species. Genes are carried on chromosomes. The biotechnologist attaches, by direct manipulation, a foreign gene possessing the required property to the chromosome of a plant or other organism. According to some scientists, such genes can be mixed, irrespective of the plant, animal or human species, without causing a problem. Depending on the 'improvement' being sought, vegetable and animal, human and goat genes are paired up. So you can find the genes of a strain of cholera bacteria in alfalfa, a chicken gene in potatoes, scorpion genes in cotton, fish in tomatoes and strawberries, firefly in fish, trout in carp, hamster in tobacco, tobacco in lettuce, and human genes in rice, tomatoes, potatoes, and ewes.

The incentive to research this field is evident: the genetic manipulation of a plant or an animal enables companies, by enforcing industrial patents, to become owners of all the modified plants and animals subsequently produced. By buying up rival seeds and patents, or removing competitors from the market, a firm can become the owner of an entire species. It's the logic of industry, applied to life. Genetic manipulation is a way of being paid royalties for life itself.

Can GMOs improve your job as farmers?

JB: We don't need GMOs to do our job. In agriculture their sole function is to deal – badly and dangerously – with the problems caused by intensive farming, especially by monoculture. Here, concentrations of parasites, insects and weeds have become resistant to pesticides. For this reason, a gene has been inserted into a chromosome of the maize plant which causes it to secrete a chemical insecticide. The maize will produce the insecticide throughout its life. If a caterpillar eats the maize, it will die on the spot.

Rape seed has been genetically modified in order to be resistant to a pesticide sold by the same company that sells the seed. This crop can therefore be sprayed copiously without damage to the 'vaccinated' plants. And single crops will continue to be grown, in ever larger areas, without any concern for the soil or the possibility of genetic pollution. And there are dangers in store!

The pesticide that has been sprayed often remains on the grain. And insects also ingest the poison. In this way, pesticides accumulate and can enter the food chain. Man is the big animal at the end of this chain. Moreover, the process of genetically modifying a plant uses antibiotic molecules as 'markers', which may promote the development of antibiotic resistance and the consequent danger to human health.

In Brazil since genetically modified soya has entered the food chain, there has been a marked increase in allergies to soya. Another worry is that laboratory rats which were fed GM potatoes are succumbing to serious immune-system disorders. Finally, there have been no studies on the accumulation in the food chain of toxic substances from GM organisms, nor on the mixing of different GM residues. Such possible 'snowball' effects have not been assessed.

Do these risks threaten wildlife?

FD: The most important risk of GMOs is the serious and irreversible effect on biodiversity brought about when they're spread in the environment. Wind and bees, the traditional carriers of pollen, spread GM pollen to traditional crops nearby, and to weeds growing in the area. Neighbouring plants could become resistant to the herbicides and pesticides synthesized by the genetically modified plants. A jungle of a new type could arise – an uncontrollable proliferation of 'mad weeds', leading eventually to the disappearance of all plant life in its natural state. A radical, uncontrolled and violent change in the world's biodiversity is on the cards. Already we've seen the ravages of GMOs on the Monarch butterfly population in the USA.

Would keeping the GM fields isolated from other crops that they might pollinate be an answer to these fears?

JB: The existing regulations envisage a 'protection perimeter' around GM fields to contain any dissemination, but that's a waste of time. The most recent tests carried out in the USA on rape seed – leaving barren strips of land fifty metres wide around the GM fields – have shown that pollination can occur at distances of up to 4.5 kilometres. Today, we're beginning to realize that the cultivation of GM crops precludes any other form of planting. In practice, genetic pollution introduced by the GM crop makes it impossible for farmers to guarantee that their harvest is free of them. Such pollution, even if it is only potential, is a terrible disadvantage for farmers who could lose their organic status if their fields are close to GM crops. We have to say no to GMOs – no other alternative is possible.

But surely these GMOs were tested before their commercial application?

JB: The big seed companies and pharmaceutical firms played on the fact that GMOs are already in the food chain, in particular through the import of soya and maize from the USA and South America, and are now unavoidable anyway.

We're convinced that if the costs of safety evaluation are included – and these are very high, as research can last from five to ten years – these products are not commercially viable. But such evaluations have never taken place. Furthermore, although 98 per cent of GMOs are plants which produce pesticides, they're not submitted to the Commission on Toxic Materials, as normal pesticides and herbicides are. The agrochemical companies have organized things so that GMOs are not classified as botanical health products, to avoid an unfavourable consumer reaction and stricter administrative controls.

FD: Despite claims that GMOs have a lifespan of ten years, they prove effective for only a short time. Parasitic plants and insects can become resistant to the 'treatment'; a different pesticide – and hence a different GMO – has to be found, or the company will lose its market. To counteract the drop in efficacy, farmers are instructed to plant a selection of non-GM plants to act as a refuge for parasites, thus slowing down their mutation rate. This is precisely what the US Minister for the Environment did in January 2000. He advised his farmers to plant less than 50 per cent of GM varieties on the same plot of land. It's grotesque!

How is the question being dealt with in France?

FD: There's a total disregard for transparency. For example, the French government refuses to tell its citizens – and, more to the point, its farmers – about any neighbouring field where GMO trials are taking place. In 1997, when the EU gave the green light for the commercial growing of the first varieties of genetically modified maize, it was the Farmers' Confederation that raised the alarm. But in fact, no precautionary measures have been taken. It's impossible even to find lists of communes where GMO trials are occurring. I'm calling on all citizens to demand that their mayor introduce a bylaw to prohibit the growing of GMOs on local land.

A TECHNIQUE OF TYRANNY

You've explained that GMOs could be patented. What problems does this patenting of living organisms pose?

FD: GM technology goes hand in hand with patenting living organisms. We're fundamentally opposed to it. From an ethical point of view, the granting of such a patent – the right to ownership of an organism capable of reproducing itself without external intervention – is tantamount to considering oneself the owner of life. This raises important philosophical and religious issues.

One of the questions raised is whether the organism, which is being modified by incorporating a foreign gene, is a human invention and, consequently, susceptible to patenting. We say categorically no. In fact, in law, a patent is applicable only for an invention that contains a new idea or technique suitable for industrial use. In the case of genetic modification, it's the technique that's suitable for patenting, not the product that results, and this is of little interest to a seed firm, although a case could be made for paying the laboratory that invented the technique.

As far as I'm concerned, patenting living organisms is fundamentally unacceptable. Life can't be patented – that's obvious. The fact that a living organism, or part of it, can now be patented is one of the biggest swindles of the century.

JB: Genetic modification is a technique of tyranny, and patenting is its main tool. It is hardly surprising that maize, the first crop to be grown on an industrial scale in the USA, was also the first plant to be genetically modified for farming purposes. When the main American seed suppliers started work on maize, they created hybrid plants incapable of producing reliable seed, and in this way they guaranteed that customers would have to go back to them every year for fresh seed supplies. That was the first stage of industrialized production.

The next stage saw the agrochemical industry invest enormous amounts to control both the sowing and the chemical treatment of crops. The five or six world chemical giants bought up the seed producers and put into practice a programme of genetic modification linked to a range of associated products: the grain is genetically modified to adapt it to a specific chemical treatment, which is sold along with it. Protected by its patent, each genetically modified seed belongs to its 'inventor', who can take legal action against any farmer who resows it. In the USA, the big seed firms employed detectives, set up special telephone lines, destroyed whole fields of crops and took farmers to court for reusing GM seeds without paying royalties.

This led to the development of self-destructive seeds using a so-called 'terminator' gene. The technique introduces a gene that stops the grain from germinating once it reaches maturity. It makes perfect commercial sense, ensuring a hundred-per-cent return on investment. The seed growers economize on the lawyers' fees that they would otherwise have incurred prosecuting farmers for theft.

These practices violate the age-old and universally recognized right of taking grain from one year's harvest to plant the next year, known as the 'farmer's sowing', which is crucial to the survival of farming communities. Until recently, this custom, recognized as 'the right to plant retention', was respected even in countries with industrialized agriculture. It's a right designed to protect the interests of those who breed new plant varieties. This system is far less totalitarian than patenting, where the seed producer or 'retainer' collects royalties on the multiplication of sowings. Once the seed has been sold, any other 'retainer' has access to them for their own purposes. Today, however, the big seed producers have abandoned this right in favour of a patent, which is much more lucrative. The abuse of patents has meant that farmers are forced to pay a fee for the use of their own seeds. For these reasons, the CNDFS, to which we belong, opposes this racket.

Under pressure from public opinion, Monsanto, one of the big players in this field, stated in 1999 that they had officially stopped the production of the 'terminator', but we have no idea what's happening to similar projects other firms are engaged in.

Does the attack on biodiversity also affect cultivation around the world?

JB: Let's look at the example of rice in Asia. There, 140,000 varieties of rice are grown for their particular properties – height, the ability to thrive in different humidities, taste or texture; wild rice, long-grain, short-grain. There are even medicinal varieties. A true rice civilization!

The multinationals are working on only five or six strains of rice, genetically modifying them for a type of intensive cultivation in areas where subsistence farming previously held sway. In some Asian countries, these five varieties now cover 60 to 70 per cent of the land planted with rice. We're witnessing the complete annihilation of a farming culture which had the ability to feed itself, together with the distinctive social and cultural system this produced. I believe this is a heavy toll to pay for this technology.

But the patenting of life is not of concern only to farmers. It has harmful effects through pirating our genetic heritage. The United Nations' Development programme has estimated the annual cost of 'bio-pirating' by the agrochemical firms at $4 billion. The major part of the planet's gene pool is in the countries of the South, but it's mainly the rich countries of the North that have the techniques and expertise to manipulate and appropriate this heritage under the protection of patents. Genetics prospectors from the multinationals scour the countries of the South in search of rare plant species that they can exploit. The example of *Margosa* in India is very illuminating. This plant has curative, insecticidal, medicinal and nutritional attributes which have made it almost a sacred plant for thousands of years. An American company had the bright idea of isolating the insecticidal element of this plant, and patented the procedures it used. The Indians themselves had isolated this element long ago, but it had never entered their heads to patent it, since they believed that the plant was public property. Today, the company owning the rights to exploit this plant could well prevent the Indian farmers from using its natural insecticide, because it competes with their own product.

AGRONOMICAL RESEARCH AND ITS MISUSE

We can see how the drive for profits encourages the biotechnology multinationals to adopt these practices. But can you explain why they're being researched in France, with public funding?

JB: French researchers have adopted the attitude: 'If we don't create GMOs, we'll be left behind the United States. If Uncle Sam is making them, there's no reason why we shouldn't.' This has been the argument of the French government – from Jospin and from ex-Minister of Education and Technology Claude Allègre – to authorize the cultivation of GMOs and encourage the development of this technology. In fact it's a very short-sighted, mercenary, and conformist viewpoint – apparently based on scientific evidence, but actually without any such thing. Now they're prevaricating, because consumers have shown that they don't want strange genes in their food.

Suddenly, our political leaders are making use of state-funded scientific research in this area. We're happy that this expertise is being used, on condition that it's completely independent of commercial interests. But can we be sure that public research is truly neutral? I believe it's generally been employed in the service of the rich and powerful.

Take the example of CIRAD, the government-financed agricultural research centre. On 5 June 1999, at the time of the 'intercontinental caravan of Indian farmers' which was travelling through Europe to protest against the multinationals, genetically modified rice being grown under glass by CIRAD in Montpellier, and due to be transplanted in the Camargue, was destroyed. In India, farmers are mounting many similar actions against the agrochemical companies engaged in GMO production. CIRAD obtained funds from Agrevo, an agrochemical multinational, to develop a rice strain that is resistant to one of the company's herbicides. This work was taken on to fund CIRAD's basic research in plant genetics. So basic research is now dependent on work with commercial applications.

Many people were shocked by the action against CIRAD. Why take it out on the scientists?

JB: We attacked their work to show that they couldn't claim to speak for independent scientific research while at the same time they worked for agribusiness. The action was also meant to challenge the way scientific research is carried out, in terms of both its principles and the source of its funding. In fact, the action provoked a debate and the publication by the collective 'GMO Watchdog' of a document in support of farmers, entitled 'Researchers, come out of your labs'.

Some researchers felt that they were under attack – I'd call that a corporatist reflex. They'd never been made to question the social value of their work. Our challenge provoked a strong response. Claude Allègre knew what he was doing when he accused us of attacking our country's independence and economy while avoiding the issue of the social role of scientific research. He put himself right in the ranks of those who own 'patents on life'.

Are you saying that research is influenced by political factors?

JB: We live under a government system which both makes the agricultural law, in the way we've already described, and promotes publicly funded research. INRA[2] supports research into agriculture on topics ranging from crops and livestock to economic and sociological aspects. It has co-operated with the modernization of agriculture; biotechnology and the state have walked hand in hand to promote their common economic interests. When biotechnology reached a level of development where it could be profitable, it was handed over to the private sector.

---

2. The National Institute for Agronomic Research, a publicly funded research establishment in Paris which has as its main function the continued improvement and development of the food industry.

A switch took place in public research, which increasingly came to service the food industries rather than the interests of ordinary farmers. The present Minister argues that public research agencies should forge links with the food and agrochemical industries; this will steer their work towards commercial interests, instead of helping farmers.

The state research organizations have forged links with commercial seed manufacturers such as Biogemma and Bioplants, and with the agrochemical branch of Rhône-Poulenc. It will be no surprise if, over the coming years, public research falls increasingly under the control of companies such as d'Avantis-Rhône-Poulenc and Limagrain. The trade unions representing workers in the French scientific community have protested about this, and stressed that these research programmes have less to do with the problems of plant genome science and more to do with the interests of the commercial market. Genoplant, an official centre for plant genome research, favours short-term technology at the expense of a multidisciplinary research programme in molecular biology that is essential to improve our understanding of plant life.

The role of Genoplant raises the question of control. What democratic guarantee is given to the public, who fund 70 per cent of the expenditure, about how these biotechnical discoveries will be used? Surely the objective of a public service must be to meet public needs, not enhance the interests of private industry.

So you believe there is a general problem with public research in agriculture?

JB: There are problems with the way research is organized and the way scientists are trained. Research projects have become more complex, and more compartmentalized rather than multidisciplinary. Biological research teams have no contact with sociologists or economists. Researchers are isolated within their fields of study and the remit of their funding contracts. They seek to promote their own work, without any awareness of how it fits into a global system.

The private firms and Chambers of Agriculture which provide the funds have a big say in the direction research takes. Plenty of work is taking place on plant diseases, on improving yield or on plant genetics, but no study is being carried out on the impact of agriculture on a world scale, or the role of research in its development. INRA should be encouraging precisely this public debate, and questioning the social objectives of the research.

There's also a problem with the training of scientists. Courses of study, at both secondary and university levels, fail to integrate the economic, social and philosophical aspects of scientific research. Students are removed from the wider culture. The great tragedy is that scientists no longer have a global view of their work. Too many have a utilitarian outlook, where the means are more important than the ends.

FEEDING THE WORLD?

Supporters of GM foods maintain that world hunger can be overcome by using GMOs to increase production, cut costs and decrease the use of chemicals. Is this so?

JB: No, this lie must be exposed. The use of GM technology is not the answer to the problem of hunger in the world. The best proof of this is that today, American farmers are suing suppliers of GM crops for loss of profits. After a few years, the performance of GM varieties becomes inferior to that of traditional strains. It's the exact opposite of the case made in the agrochemical companies' advertising. With GM crops the yield is smaller, the price of the seeds is higher, and there's no saving on crop treatments. The reliability of genetic modification is not as established as the companies make out. The technology is at the experimental stage. Moreover, no one really believes that the problems of hunger and underdevelopment can be solved by technological means; economic, social and political conditions have to be taken into account.

Let's go back to the example of GM rice in Asia, which was introduced along with the industrialization of agriculture; it has considerably weakened the farmers by reducing the varieties used, increasing the costs of sowing, preventing the resowing of grain that used to be collected freely from the fields, and failing to grow crops appropriate for particular soils.

In the Philippines, a very densely populated country, rice cultivation takes place on small plots of land, and GM genes have already polluted the traditional varieties of rice. With the aid of the International Monetary Fund, the Philippine state has set up a gene bank containing 80,000 varieties of plants, the use of which is reserved for the giant seed companies. Once the whole agricultural system has been polluted, only these companies will have access to such biodiversity. This will not be very long in coming as, currently, up to 60 per cent of intensive rice cultivations use GM strains.

What about China?

JB: This is one of the most dangerous situations. In China, the growing of GM rice has apparently gone ahead without any close control. According to the little information we have, China is the fourth biggest producer of GM crops after the USA, Argentina and Canada. Its expansion of GMO cultivation is in line with the present government's strategy to remove 250 million farmers from the land. But that raises questions about where they are to go and what they are to do.

Customers don't seem to be at all enthusiastic about GM foods . . .

JB: Increasing consumer awareness is an important economic factor. Today Japan, the biggest market for soya for human consumption, refuses to buy GM strains. Now, GM soya is cheaper than the ordinary stuff, because no one wants it.

In Europe, initial suspicion followed by outright consumer hostility have led a number of big food retailers to refuse to sell products containing GM ingredients. Consumers didn't ask for this technology, and they can see absolutely no advantage in it. Opinion polls regularly confirm this basic mistrust.

FD: Farmer and consumer mobilizations have caused a retreat on the part of the big biotechnology companies – Novartis, Monsanto, and their like. When Novartis received its first authorization to market seeds of GM maize in 1997, it anticipated cultivating 35,000 hectares in France in the first year. By the end of 1998, less than 1,200 hectares had been sown, and by the end of 1999, fewer than two hundred. But worldwide, we can count up to 400 million hectares sown with GM crops, 44 per cent more than in 1998. So the fight is just beginning.

I'm convinced that without the Farmers' Confederation and other organizations devoted to protecting the environment, we would be exposed to much greater risk of genetic pollution and dependence on agrochemistry.

We saw on the streets in Seattle that around the world, and particularly in the Southern countries, opposition to GMOs and the patenting of life runs very high. This is a very strong unifying factor in *Via campesina*. Look at the reception given to the 'Indian caravan' in France in May–June 1999. The caravan consisted of five hundred country people and farmers who came to Europe from India to spread the word about their agricultural needs, and to protest against the Western imports of soya, corn or maize.

Consumers are evidently opposed to the hidden presence of GM ingredients in their food. How can we trace these ingredients? Can labelling help?

FD: These questions preoccupy France and the rest of Europe. Supporters of GM foods oppose positive labelling (stating the GM content of the product on the label), because they're afraid that consumers would boycott such products, and because labelling would cost too much. They would prefer GM-free products to be labelled, and that the producers should bear the extra cost. This is a cynical calculation. Through lack of information, purveyors of GM produce hope to exploit the unwary customer.

Does genetic modification improve the taste or the shelf life of produce?

JB: What enhances taste is the diversity of food grown, rather than gene technology that tries to improve the taste. The use of GM technology results in a massive reduction in the variety of crops grown. In this way, it mimics intensive agriculture. For supporters of GMOs, the major concern is the yield of the crops and their suitability for transportation and processing. An opinion poll in the French press in January 2000 showed that 60 per cent of French people thought that the taste and quality of agricultural produce had deteriorated over the last ten years.

The technology for genetic modification used for plants can also be used in animals. Can this aid stock-breeding?

JB: Scientists are playing around with genetic modification without knowing the real nature of the genome. Researchers in this field admit that they know the function of only about 5 per cent of the total DNA sequences. The rest are referred to as 'junk genes'. Do we really have the expertise for safe genetic manipulation? It seems doubtful, considering that those involved don't question the possible impact of such a technology on the genomes of the full range of plant and animal life. Here we're dealing with sorcerers' apprentices who have little idea of where they're going. Because of the enormous financial stakes involved, their only aim is to find a quick technological solution. We can see this in the case of cloned animals, which are very frail and manage to survive only because they're stuffed with antibiotics.

FD: Apparently, INRA is doing research on genetically modified milk in order to produce a standardized milk so that we'll all be able to make identical cheeses. But what is the point of all Pont l'Évèques or Camemberts tasting exactly the same?

MAD COWS AND THE MADNESS OF MANKIND

The scandal of mad cow disease first broke in 1996. It exposed methods of animal feeding that the farming profession could not defend. Why were cows and other farm animals being fed animal remains?

FD: It's a logical consequence of intensive farming. When you get rid of pasture, or raise more stock than the pasture can sustain, you have to provide the animals with more concentrated food. Animals were given maize – rich in glucosides and starch, but poor in protein. Soya, which contains the necessary proteins, had to be imported from the States or Brazil, so animal feed was made from abattoir waste and the products of processing eggs, milk, meat and fish.

In order to remain competitive, the British animal feed industry was allowed to break European regulations which stated that all animal waste must be heated at above 130 degrees Centigrade for at least twenty minutes under three bars of pressure. English farmers were giving their dairy cows up to three or four kilograms of animal feed per day, in order to produce 7,000 to 10,000 kilograms of milk per cow annually. When this feed was exported to France, farmers were not aware of its contents, and had no idea that they were feeding sheep and cattle remains to ruminants, or that sick animals were being recycled. Only later did farmers discover that they had been breaking the laws of nature by making cows carnivorous, and that it was dangerous to let animals eat tissue from their own species in large quantities.

Even today, bags of animal feed do not detail the nature or the origin of their contents. The scandal of mad cow disease has been around for nearly fifteen years, and still the ingredients of feed can't be traced to its sources.

Given the lengthy incubation period for the disease, some British epidemiologists have been predicting thousands of human deaths as a result of BSE. Are we on the eve of a bigger scandal than the scandal of contaminated blood in France?

FD: The public authorities have a responsibility to guarantee to the consumer that food is safe. First in Britain, then in Brussels and France, this essential duty was ignored in the race for markets and profit.

In Britain, the disease became endemic in the mid-1980s. The Thatcher government, in its deregulatory zeal, was late in intervening, or maybe even unable to do so, as most of the animal health services had been dismantled. Breeders were left to cope with the disease, even though it was known that the risks could be eliminated by cooking animal waste destined for animal fodder at high temperatures, and stopping it being fed to cattle. Although these measures were passed in 1988, Britain did not fully implement them. Commercial pressure meant that animal feed exporters continued to sell the prohibited product throughout the EU and the Third World. In fact, the attitude adopted by the EU was not much different from that of Britain. Although the EU passed regulations covering the treatment of animal feed, the responsibility for enforcing them was devolved to member states. No central controls were put in place. Worse still – in the name of freedom of trade within the union – the EU allowed the export of British animal feed, which increased as exporters tried to compensate for the fall in demand in Britain.

France's own attitude was not much better. In 1988, BSE had already been recognized as a disease with very grave implications for human health because of its troubling similarity to Creutzfeldt-Jakob disease (CJD), a progressive degeneration of the human brain. But it was not until late 1989 that France requested that dealers in English animal feed should obtain authorization before importing the stuff. It took a further two years before these animal feeds were forbidden as fodder for herbivores.

The French state and its veterinary services adopted a 'head in the sand' policy between 1989 and 1996. Most animal feed, both English and French, was being used in clear breach of the established regulations: abattoirs were failing to cook the animal waste properly, and feed was being indiscriminately mixed in factories that also manufactured food for herbivores and other farm animals.

Some people may be reassured by the extent of the outbreak of BSE in France: only 161 cows were affected by the end of 1999, compared to 178,000 in Britain. But the length of the incubation period of CJD – from five to twenty years in humans – means that considerable uncertainties remain. In Britain, 84 people have died of CJD; in France so far there have been only a handful of reported cases.

But we musn't forget that until 1996, before the embargo, one in five steaks eaten in France came from meat imported from Britain. If some projections are accurate, there's a real danger of an explosion of the disease. According to *Le Monde*, current predictions of the number of future victims range from an optimistic 14,000 to a pessimistic half a million. So there's every reason to be concerned.

JB: The failures of the public authorities on this issue go back a long way. In summer 1996, at the height of the mad cow crisis, a debate raged in the press about the amount of British animal feed imported since 1989. A figure of more than 16,000 tonnes was mentioned, which the Ministry of Agriculture was quick to deny. It has subsequently become clear that they reported only a tenth of the actual amount of feed imported. Such duplicity by public authorities is inexcusable.

To bring attention to this falsification, we organized action at the Central Customs Service in Toulouse in September 1996. We went into the offices and made off with the Customs documentation listing all the imports of British animal feed, year by year and month by month, together with the names of the companies involved. The documents were passed to the French and European Parliaments, and to the magistrates' court in Nantes, where, as early as July the same year, the Farmers' Confederation had laid charges against fraudulent imports of animal feed.

The Customs Service did not dare sue us, even though a theft of their records had taken place. The magistrates' court has yet to prosecute a single case of the traffic in animal feed, although the documentation we provided showed how embargoed feed continued to enter the country by different routes, and which companies were involved.

The scandal of the contaminated blood products[3] soon became a criminal case. What's the situation with contaminated animal feed?

FD: In the contaminated blood affair, criminal proceedings were taken against the politicians and administrators responsible, but in the case of mad cow disease, it is only those who have to slaughter their herds who are punished.

In January 2000, the government introduced a BSE test for cattle presumed to be disease-free. The test will be carried out on about 40,000 cows in western France, the most suspect area, and it is already worrying health officials, who are afraid it will reveal a high level of infection. Their concerns are justified in the light of an investigation carried out by vets between 31 May and 15 June 1999, which revealed that EU-banned feed was still being used on French farms. There's not much point in testing for mad cow disease if the source of the contamination hasn't been eradicated.

Does this explain the breeders' silence when the disease affects their herds?

FD: With the systematic slaughtering of a herd as soon as one animal was diagnosed with BSE, the French state believed it was efficiently stopping its spread, and so reassuring customers. But for the farmers involved, this

3. In 1985, unchecked contaminated blood products were used on haemophiliacs and other patients. When it was discovered that the blood was HIV-infected, a massive cover-up, involving Cabinet ministers, was orchestrated.

arrangement was a terrible shock, especially as they were made to feel that they were to blame. A letter from the Ministry of Agriculture followed each new case of mad cow disease, asking the farmer whether there was a possible mix-up of cattle feed with other feed destined for pigs or chickens. Farmers had to ask themselves whether, over the last five or six years, some pig feed might have been given to cows.

The government maintain a blackout on the details of BSE cases, and instead put pressure on the farmer: 'Don't talk about it; don't give details; remember we pay you to have your herd put down.' In this way, farmers are placed under a vow of silence when it should be their right to take action openly against the suppliers of animal feed.

AN EMBARGO AND ITS AMBIGUITIES

Do you think the risks justified the French government's decision to put an embargo on all British beef?

JB: We don't agree with the way the embargo was imposed. Either the meat was dangerous for French consumers, and consequently must also have been so for the British – in which case a total European ban should have been implemented – or this is an example of an EU failure, where national economic rivalries damaged farmers on both sides of the Channel.

The embargo was purely political, and had nothing to do with protecting public health. To say that French consumers should be protected, but not British, was nonsense in a supposed European union. It did, however, rally farmers behind the government, and lent a helping hand to the FNSEA in building chauvinistic barricades. The FNSEA's threats to boycott English meat simply encouraged corporatism and nationalism.

What do you think should be done?

FD: The EU should take measures to ensure that all its public health directives are implemented in all countries of the Union, without exception. There must be total transparency about the channels of trade in animal feed, and when there is doubt, precaution should prevail. The Farmers' Confederation calls for the setting up of a European health agency, independent of the state, and especially of business interests.

JB: I am a representative on the European Consultative Committee on the Health and Safety of Agricultural Products. In May 1999, we called for a ban on animal feed for pigs and chickens. The animal feed industry laughed in our faces. One month later, an epidemic among Belgian chickens hit the headlines. But no one seems willing to make changes.

FD: All these health scares – mad cow disease, dioxins, growth hormones – can be laid at the door of intensive agriculture. They reinforce the need for all products to be traceable to their sources and methods of production. Nobody, consumer or farmer, should have any faith in the sales records of agricultural corporations: one will claim that they don't use animal feed; another that they don't use GMOs. But it's obvious after what's happened that you can't take them at their word.

The news that animal feed could contain sludge or oil waste was received with disgust. It seems that problems of production can be solved only by resort to the dustbin. What do you say about this bizarre way of viewing the food chain?

FD: As long as humans continue with a carnivorous diet, there will be animal waste products to be recycled or destroyed – from abattoirs, fishmongers, dairy-product manufacturers. In France, this amounts to 1,200,000 tonnes annually, which must be dealt with in one way or another.

If this mountain of waste is not recycled into animal feed, then it must be incinerated – which has its own problems.

In 1996, faced with mounting concern about the contents of the animal food chain, the Farmers' Confederation called for a ban on all animal material in fodder, because the conditions to ensure safe breeding of livestock do not exist. We still don't know today whether these conditions have improved, so we're maintaining our original cautious position.

Are you worried that these repeated food scares encourage the increasing ranks of vegetarians at the expense of the farmer?

JB: There's a great deal of iniquity in these matters. Food scares make victims of the farmers, the majority of whom are just as appalled by these scandalous practices as everyone else. The drop in the price of beef following the mad cow crisis particularly affected herds which are specially reared for meat production, and are hardly fed on maize silage or other concentrated feed at all. Mad cow disease is particularly prevalent in dairy herds. So there's a profound sense of injustice.

6

THE FACTORY FARM

THE PROBLEMS OF THE 'BRETON SYSTEM'

Poultry and pig farming have experienced repeated crises because of intensive methods. Can you describe what's happening in this sector?

FD: As early as the end of the 1960s, the emerging pig and poultry lobbies succeeded in repealing a law passed at the beginning of the decade which limited the number of pigs bred to less than one thousand per farm. Nowadays, over 2,000 sows can be bred on one farm – up to 20,000 pigs if you include piglets. What started as a complement to dairies or mixed farming has turned into an industry in its own right in less than twenty-five years.

Very quickly, poultry farming has become an integrated industrial process. A single company will sell farmers animals, feed, veterinary products and the technical information required for production. The same company will determine the type and size of building required to house the birds, which the farmer finances through borrowing. The farmer bears the running costs for things like water, electricity, heating and vet's visits, and provides his labour. The supply company decides on the repurchase price of birds for the market. The farmer controls very little: it's a pre-programmed process. He is at the mercy of the supply company and its technicians throughout the breeding and rearing process.

Consumers have increasingly spurned intensively reared and hormone-fed chickens, and turned in growing numbers to free-range produce. But industrial production has continued to flourish, especially for supplies to the catering industry and the export markets, the latter supported by French government and European grants.

A growing proportion of pig farming has also experienced this type of integration, and since the mid-1970s the sector is increasingly dominated by large producers who prioritize output at the expense of consumers' concerns and the environment.

Today, 40 per cent of pig farming is concentrated in Brittany, and some local free-market groups there have imposed a 'bid market' which is expanding throughout the region. Based in Plérin, on the Armorican coast, this twice-weekly market takes the form of a public auction between buyers and sellers. It's a flexible system that can be manipulated according to European or world markets. Some of the traders involved bet on the price variations that result. They have interests in both camps: the producers and the abattoirs. The resulting rapid price variations have forced out the smaller farmers, who are unable to borrow sufficient funds to manage periodic price slumps. In the early 1970s, there were 26,000 pig farms on the Armorican coast; now there are fewer than 3,000, but with three times the output.

JB: The big traders know how to react to the bid markets in different countries. They move their bids from one country to another according to availability and price. The system follows the same logic as the Stock Exchange. When things go wrong, members of the FNSEA will mobilize to wreck a prefecture deliberately, and seek public funds to 'disentangle' the market. They export to Eastern countries, without bothering to find out what damage they're causing to local pig farming in the countries they're trading with. The animals are sold very cheaply, well below profitable prices. It's an example of the neoliberal creed: 'Free to win and swallow up your neighbour, but also protected so that you don't lose.'

Until the early 1990s, the deficit in the French market justified an emphasis on rapidly increasing production, but the market has been punctuated every three or four years by periods of crisis when prices have slumped, followed by periods of high market prices which enable bigger breeders to make significant profits.

Concentration of production has been accompanied by a massive expansion of the processing industry, which has badly affected the market for quality home-cooked pork and salt meats from other regions, at the expense of taste and diversity. Industrially produced cooked pork and salt meats are often injected with polyphosphates that allow for the retention of water throughout the cooking process. In this way, a competitive factory owner can obtain 20 per cent more in the weight of meat going to the market. It's a foul process.

Farmers in Brittany have travelled a long way down the path of intensive production, and are often used as an example by proponents of this type of farming. But their off-soil breeding has resulted in severe environmental problems: the poisoning of groundwater, atmospheric pollution, green tides, embankment erosion, and damage to local wildlife. We should remember that the 'Breton system', praised by supporters of intensive farming, achieves its efficiency at a severe environmental cost.

FD: But the problem is not confined to Brittany, and we must be careful about making scapegoats of a single regional group. In fact, the 'Breton system' of industrial pig and poultry farming is practised all over France, even Europe, with lower levels of integration in some countries, such as Belgium, or on a few very large farms, as in Spain. Besides, pig and poultry producers are not the only cause of pollution in Brittany. The whole range of animal production pours more nitrogen on to the soil than the crops can absorb. By increasingly substituting corn silage for grass, cattle farmers have to buy more and more soya to give their animals a balanced diet. And you mustn't lump all Breton farmers together in the same category. For a long time, many small farmers in Brittany, singly but more often in groups, have resisted overproduction. Others are researching and experimenting with other methods of farming – notably André Pochon and his 'Sustainable Farming' network.[1] A 'Breton system' certainly exists, but it's not representative of most farmers there, only those with the highest outputs.

These qualifications notwithstanding, the situation in Brittany is certainly serious, and illustrates the dead-end of overproduction as promoted by joint management between the state and the major agricultural organizations.

In Brittany, total farm animal excrements are equivalent to a human population of 35,000 million inhabitants. A healthy pig, for instance, discharges seven litres of faeces and urine every day. Normally, animal faeces are a good fertilizer for cultivated land: manure, a fermented mixture of faeces and straw, warms up, enriches and airs the soil; but excrement produced by animals in industrial buildings, housed not on straw but on duckboard, is more difficult to use. The composition of animal excrement in industrial farming differs from normal farming: it's smelly, laden with heavy metals, and often contains undigested residues of antibiotics.

The volume of excrement released is such that the area of land available to absorb it is insufficient, so that, between mineral fertilizers and animal faeces, Breton farmers spread 30 per cent more nitrogen than necessary on cultivated land. As a result, thousands of tons of nitrogen are released into the air (as ammonia) and into the water system (as nitrate).

This environmental disaster affects all activities in the region; consumers are forced to buy spring water for cooking and drinking. Breton springs, such as Katell Roc, have had to close down their pumping facilities because of contamination. Surface and underground water pollution is such that a report in 1995, drawn up at the request of local groups, indicates that if present-day farming trends persist, Brittany will have to close down

1. The 'Sustainable Farming' network organizes farmers involved in the struggle for economically viable and autonomous farming. See Appendix 2.

three-quarters of its drinking water supply network to meet the European norm of 50 mg of nitrates per litre. And that's not a particularly severe restriction – the World Health Organization strongly discourages babies and pregnant women from drinking water that contains only 25 mg of nitrates per litre. The poor quality of water also affects local industry: fishing and fish farming have suffered, and tourism is hit because of polluted beaches. Local food-processing companies find it increasingly difficult to obtain good-quality water for their factories.

But people are on the move: many pressure groups have been set up in Brittany to attempt to protect the water system. Have you worked with these groups?

FD: Fishermen were the first to react. Sunday fishermen are just regular citizens for the rest of the week, and come from many different social and professional backgrounds. This has given their protests particular weight. The association they set up, 'Waters and Rivers of Brittany', has become a major environmental lobby. It knows how to motivate ordinary people, other groups, and local politicians. Numerous other environmental organizations, such as the 'Coherence' network and the 'Pure Water' collective, of which we are members, are attempting to mobilize the Breton population. Individual farmers, the Farmers' Confederation and also the agrobiologists' federations and 'Sustainable Farming' are working alongside consumers to protect the quality of water and promote farming that is in harmony with the land. On 21 March 1999, in Pontivy, out of seven thousand demonstrators, there were close to one thousand farmers demanding that water supplies should be protected, and that there should be a new farming policy for Brittany.

LOBBYING AND INFLUENCE

Farming off-soil, especially pig farming, is very often condemned by environmental groups as against the law. Why?

FD: Over the past twenty-five years, producers' associations have systematically encouraged their farmers to disobey the law on 'registered sites': locations where cattle breeding is supposed to conform to norms for the protection of the environment. Over 30 per cent of new buildings and extensions were constructed without permission, or without the installation of appropriate disposal systems for animal excrement.

The biggest farms have led the way, illegally building extensions three times over the allowed size. Getting around the law is a popular sport, which bankers never seem to balk at financing, and which flourishes due to lax – even obliging – public authorities. With so few controls in place, it's obvious that officials very often turn a blind eye. Complaints from a neighbour or an environmental group are often filed in such a way that there's no official follow-up.

When local people object to such violations, they're blackmailed with arguments which insist that their actions will threaten jobs, the balance of trade, the region's status as exporter, and so on. And if farmers dare to complain, they're met with thinly disguised threats of economic boycotts and, in some cases, physical attacks against their farms.

In the past few years, successive governments have taken measures to control agricultural pollution. How successful has this been?

JB: The European directive on the reduction of agricultural nitrates in surface and underground waters had to be passed before the major professional organizations accepted, in October 1993, the need to negotiate a five-year plan on the control of agricultural pollution with Édouard Balladur's government. The PMPOA[2] made the control of water quality their chief objective.

The professional organizations' primary objective was to obtain very high subsidies: 65 per cent from public funds, to cover the cost of the work and equipment needed to dispose of the waste. These subsidies were available only to the large breeders, because they were based on the number of animals being farmed. Member organizations, familiar with joint management, have secured the right for their farmers who are benefiting from the programme to be subjected to the water authorities'[3] pollution tax only after the work has been completed. This is a very privileged arrangement in comparison to other industries, which have been subject to the same tax for many years. It is contrary to the 'polluter pays' principle. The FNSEA has declared itself ready to take environmental protection issues into account, but at a cost acceptable to farmers. It has attempted to pass on the cost of treating and repairing damage to the environment caused by farms to the community as a whole.

FD: Five years after its launch, PMPOA's inadequacies are clearly exposed, and the FNSEA's double-speak has not convinced public opinion. An interdepartmental report called 'Agriculture, Environment and Budget', drawn up in autumn 1999, which the government tried to keep secret in order not to upset the FNSEA, is a severe indictment of the programme, its objectives and its enforcement. In relation to soil testing, the authors actively question the programme's capacity to guarantee water purity. It's a financial black hole which has swallowed up 10 billion francs' worth of public money. The report underlines the adverse effects of a system which favours the biggest farmers by allowing subsidies without upper limits. The investigators also exposed the allocation of public subsidies to farms with numbers of

2. PMPOA: a five-year programme designed to abolish pollutants of agricultural origin.

3. The pollution tax is calculated according to the levels of pollution caused by agricultural activity contaminating water.

animals beyond the legal limit. They showed how the programme was distorted so that it actually encouraged farmers to break the law in order to get subsidies for the modernization of their farms. We're in a situation where local authorities have to pay farmers merely to get them to obey the law. It's an environmental fiasco, a political and financial scandal that raises questions about the entire strategy of joint management in agricultural policy.

As early as 1996, we in the Farmers' Confederation expressed exactly the same criticisms as the interdepartmental report. We pointed to the discrimination against small and medium-sized farmers, emphasizing that, in the long run, the policy threatened their very survival.

Statistics indicate that only 20 per cent of Breton farmers have changed their methods. Is their lobby too powerful, or is there a lack of political will to enforce the law?

FD: PMPOA has failed because it doesn't deal with the basic problem of the concentration of pollution per hectare. Digging bigger sewage pits can restrict the length of time when manure is spread on the soil, but it just results in more intensive spreading over a shorter period – it doesn't affect the volume of manure going on to the land. It simply doesn't make sense to subsidize the construction of bigger sewage pits by up to 65 per cent of public funds. If numbers of animals are not reduced, the result is the same excess of pollution. Beyond a certain size, these farms are industrial enterprises, and need to develop industrial solutions to the treatment of sewage, taking responsibility for the associated costs. They can't be allowed to ruin good land at the expense of small and average-sized farms.

It goes without saying that this is not the solution proposed by pig and poultry farming lobbies, nor the one imposed by the authorities. The current arrangements simply result in larger farms pouncing on their bankrupt, smaller neighbours in order to expand their holdings.

Those big farmers' lobbies have fought hard with the European Commission to ensure the continuation of corn subsidies, so that they can sow and feed corn which absorbs large amounts of manure. If corn, which is greedy for nitrates, were not subsidized in this way, if these areas were returned to their original state, there would be severe air-quality problems resulting from the spreading of surface manure. People would rebel. From the big farmers' point of view, it has become imperative to continue intensive ploughing as the only way to bury large volumes of manure in small areas. But because of twice-yearly ploughing, the soil loses its retentive properties and no longer acts as a filter. So nowadays, an interregional system of sewage pits has been developed. Lorries filled with poultry droppings from the Armorican coast drop their cargo on the cereal-farming plains around Paris. The system is out of control and running ahead of itself, but it's maintained by the European Union, which pays 2,800 francs in subsidies per hectare of corn fodder, but finds it hard to pay 300 francs in subsidies per hectare of grass.

How profitable is this system overall? The costs of concentrated industrial production are, in the end, borne by local farming communities, and the future of those regions with water or tourist problems has been mortgaged for decades to come. The European Union, the state, the region, the Department – all contribute to these anti-pollution financial support initiatives. Taxpayers have to fork out, but the root of the problem is not addressed.

Is the French example an exception in Europe?

JB: No. Both the Netherlands and France delayed implementing the European Union 'nitrate directive' until 1992. In the Netherlands, the situation was even worse than it was in Brittany. The concentration of animals was higher, on even more porous land. But subsequently, the authorities there took advantage of the damage caused by swine fever, which decimated close to 25 per cent of pigs between 1994 and 1996, and imposed a drastic 20 per cent drop in production.

ESCAPING FROM OVERPRODUCTION

On top of this environmental deadlock, pig and poultry farmers have experienced an unprecedented crisis over the past several months. Why?

FD: Different factors are involved in the pig and poultry sectors. In the case of pig-rearing, profound changes in production methods are connected to overproduction. When market prices remain below 7 francs per kilo, as they have for two years now, very few producers can earn a living. But production costs vary from one farm to another, depending on the level of debt, technical expertise, and the size of the farm. Even successful small farmers can't survive for very long. On the other hand, big producers are less vulnerable because they benefit from preferential treatment within their own groups and co-operatives. For instance, discounts are available when large quantities of feed are purchased. These discounts can be significant and, as a result, the cost of production varies from one farm to the next.

All farmers get into debt or draw on savings built up when prices are higher. The overall level of debt in the industry is astronomical, and it affects not only breeders but the whole of farming. It's primarily a financial crisis. Since the production system has already been modernized, there are few small farms left to be scrapped. The farmer's individual survival is linked to economic performance, but it's also 'managed' by his partners. There's no room for breeders whom the banks or producers' associations are unwilling to help. The selection process determining who will survive and who will go to the wall operates according to economic criteria. But other considerations also come into the picture. The attitude of bankers and associations is important, and experience has shown that they are unlikely to favour, for instance, those who are active trade-union members.

When farmers are forced to stop trading because their debt exceeds the value of their farm, they're generally soon taken over by other breeders. Their livestock is not sent to the abattoir; previous production levels are maintained. Occasionally the bankrupt farmer is kept on as hired help for the new owner, usually under much worse working conditions. This brutal consolidation is at the root of a crisis in pig production that has been acute for over two years now. Producers' associations and feed suppliers are often actively involved in the takeover of bankrupt farms, using their financial muscle to streamline their production in a way that makes them more competitive. Of course, the costs they incur in implementing these changes are subsequently unloaded on the remaining farmers. And the competition involved here is not just at the co-operative, or even regional, level. It's international, with the four major areas of production – Brittany, Holland, Denmark and Spain – battling it out against one another.

FD: The pig crisis is now so acute that it can't be rectified by simply slaughtering the livestock of bankrupt farmers. France and Europe in general are more than self-sufficient in pig production: we've reached 107–108 per cent of our needs, an enormous surplus. Yet despite this, no policy to control production has been put in place. Politicians attempt to reassure producers by pointing to the potential of the world market. But this is an illusion: world trade represents only 3 per cent of total production. Moreover, European pig farming networks can compete on the world market only with major subsidies, inevitably financed by the taxpayer.

As far as poultry farming is concerned, the same production-driven logic prevails, but the breeders' situation is different. As I've already pointed out, production in this sector has been controlled for a long time by a single organization that owns a large majority of the feed manufacturers, the abattoirs and the livestock.

Poultry farming today is suffering the consequences of the Marrakesh agreements signed in 1994, which liberalized trade and removed export regulations. Without the backing of the European Union, the big names in French and European poultry farming, such as Doux or Bourgoin, are no longer competitive. They've been forced to move their production to places like Brazil and Hungary, where they enjoy cheaper labour and lower feed prices, and can benefit from the devaluation of local currencies. We've seen similar developments in the textile and shoe industries, and in household appliance manufacturing.

What solutions do you suggest to escape this dead-end in intensive farming?

FD: First, we must control production. Unfortunately, the debate about what needs to be done reaches the general public only when the crisis is already so deep that farmers don't want to hear proposals for cutting output. They feel trapped, especially since they have been told for the past thirty years that the only way forward is to increase production. For us, control of production can't be separated from regulation of farm size. But if production were concentrated on the internal market, export allowances could be progressively reduced, and consumer demand for quality and environmental concerns could be given greater weight.

JB: The structural crisis is extremely serious, and demands a commitment from French and European public authorities. Transforming the system is by no means straightforward, because intensive farming is basically a heavy industry. It's hard to change an industrial pig farm into a different sort of agriculture. But we need to get rid of the off-soil model and replace it with alternative methods. I favour a plan for total reorganization, spread over several years. Thousands of jobs in agriculture and the farm-produce industry are concerned, and it will take time. We must start from the internal needs of the European Union and get rid of overproduction, by reducing output on larger farms, until we get back to a level that does not exceed national self-sufficiency. A plan should then be introduced to lower the level of production progressively in some areas, spreading it more evenly across the country.

Subsidies should be granted only to farmers setting up non-polluting, small projects. But it can't happen overnight. You can't eliminate forty years of industrial folly in a couple of years, and this sort of change is going to take enormous political will. And it can't be achieved just on a national basis; it has to be worked at on a European level.

But how can this strategy prevail, if politicians refuse to act?

FD: Pressure from environmental organizations can help to change the rules of the game. Associations such as 'Waters and Rivers of Brittany', fishermen's associations, ecologists, naturalists, people who are simply fed up with buying bottled water – in short, action by people who decide to try to take control of their own environment, and are numerous enough to put pressure on politicians. In this struggle, rural people are supported by city-dwellers with an interest in the country.

And what about practical measures to help farmers caught up in the current crisis?

FD: We have to ensure that small and average-sized pig farmers are targeted for public aid. We're lobbying to force the pig industry to provide debt relief for struggling farmers based on clear, public criteria and managed by *ad hoc* commissions. Since the beginning of the crisis, we've been asking the government to force the biggest producers to slaughter the sows which they keep in excess of legal limits; but until now, public authorities have done very little about this – political lobbying has been too powerful. The government upholds rights, but it's obvious that some have more rights than others. . . .

As for industrial poultry farmers, the primary requirement is for a renegotiation of their contracts. Everything else depends on this: work conditions, remuneration, renewal of livestock, construction of farm buildings. They're in an unusual situation: they're both home-workers and part of the system. We're lobbying for equal rights when contracts are negotiated. But this crisis will not be resolved without a restructuring plan. We argue that such a plan must unite the government with the private companies in order to implement a sweeping reorganization of the industry that is acceptable to breeders.

PART THREE

WE CAN CHANGE
THE WORLD

7

SHARING THE LAND

The countryside doesn't belong only to the farmer: other people live there, often looking for a better quality of life than they had in the cities. They regularly run up against legislation specifically geared towards agriculture. For a long time – since 1932, in fact – the French urban population has been larger than its rural counterpart. City-dwellers are victims of the rush of urban life, of pollution and stress, and – quite reasonably – demand the right to go and breathe fresh air in the countryside every now and then. The food scares have spurred them on to find out for themselves what exactly is on their plates, leading them to the farmyards. Farmers, country-dwellers and town residents are forging new social links, and redrawing the boundaries in regions where private plots and collective land coexist. We are witnessing the redefinition of the farmer's job. The way this shakes out will determine whether the farmer remains pivotal to the organization of land, or becomes simply a bit-player in a huge leisure park.

GOING BACK TO MULTIFUNCTIONALISM

Before the agricultural specialization you've described, farmers worked a multicrop and multibreeding system. According to the seasons, they would sow, reap, repair their equipment and buildings, tend the woods and trim the hedges, move the animals on to different land. They would also cultivate the kitchen garden and look after poultry. In short, they were, of necessity, multiskilled. Since the last CAP reform, a new word is being used: 'multifunctionalism'. What is this?

JB: We don't believe the farmer's job can be reduced to marketing. Farmers work with what's alive and on the land. They provide employment, help to conserve biodiversity, and preserve and maintain the countryside. Choices made by farmers directly affect the land and the environment. There are three dimensions to the farmer's job: economic, social, and environmental; and it's the integration of these three – or lack of it – that defines agriculture today. Previously, farming was a rather closed world, but today, developments in research and the impact of different experiences and knowledge, together with society's demands for good-quality food and general environmental concerns, have forced farmers into helping to create a different and better future.

We believe that such diverse responsibilities must be taken into account in deciding how much a farmer's job is worth. Only when their income is guaranteed by fair prices will the farmers be able to live harmoniously, and take on the non-commercial functions of the job. It's the farmer's produce that pays for all these non-commercial aspects, so there should be no distinction between productive work, and environmental and social responsibilities.

FD: For a number of years, we've seen the concept of 'multifunctionalism' introduced in agricultural policy documents both at the European and the French level, recognizing that agriculture fulfils not one function, but many functions, some of which are not commercially viable. This was shown by the recent agricultural directive from Louis Le Pensec, Minister of Agriculture, adopted by the French Parliament in June 1999. This directive assigns economic, social and environmental responsibilities to agriculture. It also officially recognizes the failure of the intensive system of farming. That initiative was a real breakthrough.

Multifunctionalism directly echoes the slogan, 'Produce, Employ and Preserve', our farming triptych. But many questions remain unanswered. In the first place, not everyone puts the same weight on the word multifunctionalism. For us, it is a path towards an agriculture that shows more respect for people, the soil and animals; for others, it's merely a pretext to get extra bonuses.

Are you saying that there are different interpretations of multifunctionalism?

FD: Although multifunctionalism has been written into the law, the Minister of Agriculture has not given up his support for intensive farming. There is the risk of a confused interpretation of multifunctionalism. The main danger we fear is a token endorsement of multifunctionalism without an overall strategy for the technical and economic choices that face agriculture.

JB: Multifunctionalism reflects a balance of forces. The EU acknowledges the strength of consumer feeling in favour of wholesome food, consumers' respect for biodiversity and their concerns about factory farming. We in the Farmers' Confederation don't want multifunctionalism to become a way of accepting intensive agriculture by giving society the impression that the state is taking care of the countryside. We don't want a handful of farmers providing a cover of rustic authenticity by looking after a few hedges, flowers and birds – a sort of cardboard-cutout countryside.

Our analysis of government budgets leads us to believe that the agricultural directive was not interpreted coherently; public funding of French agriculture is close to £10 billion annually, of which £7 billion is in the form of direct grants. Virtually all the money goes to intensive agriculture, whereas the amount allocated to farming under CTE contracts is only £0.2 billion. Enough to fuel our worries. . . .

What's CTE?

FD: The Contract for Territorial Exploitation was recently created by a new farming law designed to encourage multifunctionalism in agriculture. A CTE is a five-year contract between the state and a farmer. In exchange for financial aid, farmers commit themselves to apply measures in two aspects of farming: a socioeconomic one (promoting food quality, job protection and agricultural diversification) and an environmental one (encouraging reduced fertilizer use and the fight against soil erosion). The individual farmer's commitment is laid out in a contract that is negotiated between the state and professional agricultural organizations at a Departmental level under the control of the CDOA.

The CTE is simply a tool – a lot depends on how it's used, on what objectives one wants to achieve. We like to think of the CTE as an instrument that enables agriculture to move away from its present emphasis on increasing productivity to one that maintains, or even increases, the number of agricultural jobs.

From the standpoint of the Ministry of Agriculture, the CTE is a way of encouraging the protection of the environment – farmers will get more subsidies if they adopt measures that respect the environment. That sounds fine in theory, but in practice it has meant that those who already have access to funds now receive more money to pollute a little less, but stay within the system. Such farmers take the maize premiums with one hand, and with the other grab the grants for sowing grass along riverbanks and planting windbreak hedges around fields of GM crops, or industrial pigsties. The result is a two-tier agriculture: a lower tier of intensive but 'landscaped' farming, producing cheap standardized food for the poor; and a tier of sustainable agriculture, providing quality farm-fresh food for the well-off.

JB: To expand on what François has just said – we must stress that the CTE is an optional contract, but it prevents the emergence of a clear land use policy. In agriculture, changing practices on one plot of land while continuing to pollute the one next door doesn't provide an alternative. Even within a single commune, farms will continue to pursue conflicting agricultural practices.

DIVERSITY AND ENTERPRISE

Recently we've seen farmers diversifying their activities; they run hotels or restaurants, sell farm produce direct to the public, or organize farm visits.

FD: Yes, it's true that for a number of years now, farmers have got involved in activities that complement their job. It first started with those farmers who had rejected intensive agriculture and developed a new farmer's economy, based on a direct relationship with the consumer. This broadening of horizons took various forms: direct sales of farm produce through local markets and the Internet, agri-tourism with bed-and-breakfast, camping facilities, farm visits with guided walks, and educational activities. Such ventures have contributed to the emergence of new social networks in the countryside, which you describe very well in your book *Les Nouveaux Paysans* (The New Farmers).

The FNSEA, having spent a long time denigrating and even attempting to suppress these activities, has now changed tack. When it realized that consumers were rushing for these alternative ventures, it quickly joined in. True to form, it tried to industrialize these innovative practices and, in the process, grab some public funding. That's why today we see a proliferation of *chambres d'hôtes*. Initially these existed for the purpose of meeting people, and sharing rural experiences; but under pressure from the intensive farming lobby, they've become like a hotel chain.

JB: The multiskilled farmer is not a new concept. It's an old tradition in a number of regions, including the Alps, the Pyrenees, and the Cévennes. The wine-growing areas in the Midi have long experience of this multifunctional approach, and it continues to this day. The farmers held down two different jobs, each with a different status: on one hand, they were farmers; on the other, wage-earners. It was an economic necessity. In spring, summer and autumn they would farm, and in winter they would pursue another activity at home or in another region. In the past it was watchmaker, chimneysweep, wooden-toy maker, but today it's more likely to be ski instructor or some other job in the winter tourist industry. Monoculture was a 1960s invention; it's not traditional farming practice. In order to make farmers specialize in single crop production, full-time 'experts' were introduced who – as we've already seen – kept real farming practice at arm's length. There's nothing wrong with farmers supplementing their work, but what is paradoxical is that this practice should ever have been denigrated. Today, there's a danger that such a diversity of enterprise will be used to plug the many gaps in a rural economy that has been devastated by intensive agriculture.

FD: The business of providing farm holidays and study trips has evolved because it meets an increasing demand from town-dwellers and non-farming country folk. I think it's all part of the consumer's desire to acquire a better knowledge of the quality of their food. The consumer is trying to re-establish contact with nature, and with the men and women who work on the land.

Town-dwellers like the countryside, the feel of the outdoors, the social life of farming communities; they don't want to visit an industrialized breeding farm with a shed of 2,000 sows, or a battery farm with 30,000 laying hens. The farm holiday business can thrive only in the context of traditional agriculture.

What do you think about these 'teaching farms' which educate the public about the variety of agricultural activities?

JB: I think it's a good idea if they're real active farms, otherwise they become like zoos or museums. A teaching farm allows children who've never seen farm animals to touch and feel them, and to learn how a farm works. Just outside Millau, there's a farmer who has set up a teaching farm where he welcomes children from primary schools. This has allowed him to save his farm, with enough work for his son to consider joining him.

FD: The danger would be if, in an ocean of intensive agriculture, we saw the appearance of a few small islands of teaching farms that function only as museums. The example José has just referred to illustrates the Farmers' Confederation attitude to such activities. They mustn't be seen as a substitute for real agriculture; nor should they be merely decorative fronts, only pretending to be real farms, like some of the *auberge* farms that can't grow the 'home produce' they sell. These are farms only in name. The law authorizes such educational ventures on condition that they take place on the farm. But its wording is too vague. It's our belief that these activities should always be secondary to real agricultural work.

JB: I'm convinced that you can't fool the public for long: visitors ask questions that farmers on intensive farms have to answer. They have to explain their methods when questions are raised: 'What do you add to your products?' or 'What's that seasoning or colouring?' People don't accept any old story just because it comes from a farmer; they know about food quality, health issues, and industrialized agriculture. And that's how it should be.

What do you think of the direct selling of farm produce?

FD: It's diversification. More often than not, it enables farmers to reappropriate the tasks and knowledge they had before production methods were intensified. It's a natural extension of their jobs. That's how it is on José's farm, where he processes approximately half his ewe's milk into cheese himself. Consumers seem to like it – look at their passion for farmers' produce.

But when these supplementary activities start to take over services such as the local Post Office, snow-ploughing, trimming roadside hedges, or keeping up the communal land reserves, isn't there a risk of encroaching on people's livelihoods?

JB: There's no ready-made answer to the sharing of human work in the countryside. For example, where I live, with only two of us occupying one square kilometre, services like snow-clearing are contracted through our CUMA on behalf of communes. A single commune can't afford to buy a snow-plough, nor can it clear hundreds of kilometres of road. But by spreading the cost of the operation over all the communes, we can afford a plough and a driver, and get this job done.

MAKING GOOD USE OF THE LAND

France has traditionally relied on rural planning to develop the countryside. But intensive farming, and the monoculture that goes with it, have skewed the effects of these plans. Are farmers still in charge?

FD: Farmers make up barely 5 per cent of the working population, but despite being a minority even in the rural areas, they control and manage the countryside through their agricultural institutions. This prerogative is already being eroded by the development of green tourism, fishing, hunting and rambling, which can come into conflict with farmers' interests.

JB: There's a fundamental difference between what has happened during the last forty years and the farmers' communal traditions. Their demands for access to the land have generally had egalitarian motives. But the legacy of the French Revolution and the second half of the twentieth century have given rise to the individualist farmer, devoted to relentlessly increasing output. This new attitude towards the land and production is in complete contrast with the history of the farmers' movement wherever in the world you look. In all the big historical struggles for land – in France, in Russia, in

Mexico, or in Brazil today – farmers have shown a willingness to redistribute land equally, in accordance with the number of workers in a family or the capacity of each individual to work the land.

The 'cahiers de doléances', the system that prevailed before the 1789 Estates General, expressed this desire. By abolishing the division of the land into two parts – one for the landlord (known as the 'eminent' domain) and one for the use by the community (known as the 'used' domain) – the French Revolution created a system whereby all the land was exploited by the communities. But the Napoleonic Civil Code changed all this back with the introduction of individual property rights. One-and-a-half centuries later, the agricultural laws of 1960 and 1962 encouraged farmers to buy land by providing very attractive loan facilities, and stipulating that they had first option in the purchase of the land they cultivated.

FD: The Farmer's Statute, which became law in 1946, was a major gain for farmers. It offered them protection and security from landowners. It was a big relief for a large number of farmers and leaseholders, especially in those areas where a lot of small farms were owned by the landed gentry. Today, after fifty years of modernization, things have changed everywhere: farms have become bigger, and landowners have scattered rather than concentrated their holdings.

But the Farmer's Statute remains indispensable: it's a way for the farmer to avoid getting into debt when he's buying land and equipment.

JB: In the Larzac, we've adopted another approach, with a unique fifteen-year experiment in collectively managing 6,300 hectares of land, spread over five cantons. When the project to extend the military camp was ditched in 1981, 6,300 hectares became available for development. These plots of land, close together and free from private ownership, were seen by the farmers as having huge potential for agriculture. Our first priority was to set up an 'Installation Commission', to regulate the collective management of the plots. The Commission established criteria for choosing tenants for these plots, with preference being given to those proposing useful projects or requiring a significant workforce – our priority was to expand the farming population and develop the community.

We then established a 'Commission for the Development of the Larzac Land Project', with representation from all the plateau's 112 communes. Meeting once a month over the first three years, this Commission drew up an inventory of all the available land, and allocated it to the local farmers. It also took responsibility for buildings not used for agriculture. The distribution of land took into account those farmers who held a lease before the army expropriated it – 32,000 hectares were returned to former tenants. A further 2,800 hectares were divided between twenty-two new ventures, including seven that had already been set up during the struggle and fifteen new ones. Also, to help young farmers settle in, some established farmers gave up plots of land, without compensation. A lot of land changed hands so that it was more efficiently distributed around the main farm buildings. It took us barely three years to sort all this out.

To set up a management structure independent of the state, we devised an original legal solution based on a company of associates, which has the responsibility for looking after both buildings and land in the Larzac. On 29 April 1985, Larzac Land Associates (SCTL) signed a long-term contract with the state that is renewable after sixty years. The membership of the management council reflected the majority of agricultural workers, but also included some non-agricultural representatives, because we wanted to promote the development of non-farming activities so that life on the Larzac was properly balanced.

What is the nature of the leases or contracts between the SCTL and the users of the land?

JB: The SCTL offers three sorts of contract. For farmers there's the 'job lease', which guarantees the use of the farm for the duration of the farmer's professional life, and expires when he retires. It's a normal agricultural lease which can't be passed on to descendants. Like all rural leases, the price is fixed according to the prefectural scale, based on a 'points' system. In France, less than a half of one per cent of farmers benefit from this arrangement, because private landlords refuse to enter into contracts longer than nine years.

For non-farmers, there's a 'user's lease', which specifies how the property is to be used: for example, workshop, retail shop or home. Its duration is fixed by the SCTL according to the tenant's professional lifespan. It's not transferable to offspring, and it's free. The SCTL preferred this type of lease to rent because it got round the issue of 'key money'. The demand for a monetary deposit excluded some people because of the cost; as a result, the collective nature of the venture would be threatened. Since a lease can't be refused to someone who has the deposit, the demand for 'key money' would mean that the SCTL could no longer choose its tenants.

Another issue preoccupying the SCTL from the outset was the pitiful state of many of the buildings; rehabilitating the farms and houses required serious money, which the SCTL didn't have. It was left to the occupants to take charge of repairs, for which they were reimbursed. The SCTL was determined to ensure that anyone who left was fairly compensated and anyone who retired had sufficient deposit to live elsewhere. Finally, a third type of contract gave relevant local organizations the right to hunt in the area on land which was previously private, and had subsequently been sold to the army.

What lessons can you draw from this experience that could be applied elsewhere?

JB: Not only that it is possible to run a large estate collectively — many French communes are smaller — but that it can be a success right from the start, provided it is organized around the clearly defined interests of its users.

The ability to set up home without having to purchase work tools was decisive. It was important to reintroduce a non-mercantile element, and ensure that land was no longer subject to market forces. This doesn't mean that private property has to be abolished, but it does mean that its management must be controlled collectively on the level of a commune, or even a canton.

Access to the land must be separated from the deeds to the property. Owners' obligation to lease their land should be written into the deeds. Any landowner refusing such an obligation should have their land compulsorily leased, with priority going to young new farmers rather than adding it to existing farms. If the political will is there, these projects are realizable. They don't even challenge inheritance rights, or the existence of deeds to the property.

So it's the implementation of the principle that 'land belongs to those who work it'?

JB: No, not exclusively. It's not a question of land for those who work it but, rather, land for the collective use of the local inhabitants. Communal use of the land is what's important – the collective interest must prevail. So it's not necessarily those who work the land who decide. The people who live in an area have to decide how its resources are to be used.

But doesn't this create problems in choosing the methods of farming?

JB: We'd have to adhere to municipal advice, and agricultural constructions on the land would be introduced on the basis of a scale of charges which would be mandatory and non-negotiable. For example, organic farming needs more land, and for this reason the expansion of such a farm should take precedence over the creation of new ones. We fought for these principles in 1982–83, and came to blows over them with the Socialist government of the time.

FD: I must say at this point that the Socialist Party's record here has been catastrophic. Between 1987 and 1992, 72 per cent of land freed up by generous funding for early retirement went towards the systematic expansion of intensive agriculture.

PASSING ON ASSETS, KNOWLEDGE, OR DEBTS?

You've often stressed a lack of career opportunities under intensive farming. Where are we at, in France, when it comes to the renewal of agricultural employment?

FD: The figures speak for themselves: in 1999, there were about 12,000 new or reoccupied farms, of which 5,000 received no state aid. Against this, in the same period there were 50,000 closures – an 80 per cent deficit!

This haemorrhage can be attributed mainly to the industrialization process. After forty years of intensive agriculture and farm mergers, agricultural production demands far more capital than an individual farmer can raise.

JB: Obviously there's a limit to the level of debt one farmer can support. For a young person who has nothing, it's impossible to borrow sufficient funds to buy a cereal farm valued at two to three million francs ($300,000–$400,000), or a milk production unit at 700,000 to 800,000 ($90,000–$100,000). The income wouldn't feed the family and pay back the debts, even if one partner works outside and brings home a separate salary.

On the other hand, on our GAEC farm, the latest recruit, a young man named Nicolas Pecrix, raised 300,000 francs to buy the share of someone who was leaving. He borrowed 200,000 francs ($30,000) from the state scheme for young farmers and 100,000 francs ($15,000) as a personal loan. This isn't a colossal investment, but it immediately brings in a monthly net income of 7,500 francs.

FD: As in every other sector of the economy, the prevailing system exacerbates the inequalities between farmers. Some farms function both economically and legally like companies, with one or more managers. Sometimes – as is the case among the Bordeaux or Champagnois wine producers – the manager has limited power, with shareholders calling the shots. At the other end of the spectrum, as we saw in the intensive chicken and pork sectors, farmers no longer own the means of production. They've effectively become home-workers.

It's a two-pronged crisis: on the one hand, most modernized farms are valued at such a high price that young people can't afford them unless they're the only sons of rich farmers. On the other hand, they can't get loans for modestly sized farms which are judged 'non-competitive'. These farms are destined either to go towards enlarging the big farms or to be purchased without the aid of public funds – something a farmer's son is not in a position to do.

Aren't there viable, average-sized farms, perfect for taking over, which end up being incorporated into neighbouring farms for lack of a buyer?

FD: Today, fewer and fewer sons and daughters of farmers choose to run the family farm. Yet paradoxically, agricultural schools have never had so many students from urban or non-farming rural backgrounds. Unfortunately, state aid, in the form of preferential loans, is available only to children of farmers. Concessions are not available to someone from outside farming; that makes it financially impossible for others to enter the industry.

Most young newcomers to farming go through the GAEC or a company; group agriculture provides an ideal basis for lightening the financial burden. However, many of these ventures are purely financial and economic operations, with very little in common with the principle of sharing work. Young people are taken on only when they can get subsidies, or because they're cheaper than other workers.

Is it a good or a bad thing when external capital is put into farms? Doesn't this speed up intensive farming?

JB: Indeed it does. The big suppliers and sales co-operatives want to maintain their trade with the large agricultural producers, but when the latter can no longer survive, they will concoct some financial deal to 'support' the young farmer. In reality, it's to maintain their own income. At present, they find that the easiest solution is to offer subsidies in exchange for a commitment to remain their customer for several years.

The less intensively run farms, with less capital per head, will produce a higher rate of profit than the equivalent intensive farm when they're farmed correctly. François's farm is an excellent example of an autonomous and financially viable farm.

Of the 12,000 farms set up in 1999, 5,000 received no public aid. How come almost half of those starting up as farmers have no right to public funding?

FD: These farms are usually set up outside intensive farming, with projects that are not welcomed by the Departmental commissioners, who control the allocation of start-up money. Fortunately, some areas do give subsidies to these projects, albeit on a much smaller scale than government subsidies. We're demanding proper recognition for all such farms, and the provision of funds for setting up new ones.

But given the shortage of genuine vocations among farmers, the profession will inevitably revitalize itself from the cities.

JB: If agricultural employment is to be kept at the present level, it can be achieved only if people come into it from outside the farming sector. Nowadays the desire to set up in the countryside and become a farmer draws its appeal from the prospect of an alternative lifestyle. These young and not-so-young 'neo-rurals' who move to the land do so with different aspirations, rejecting current thinking based on markets and money. Their farms are often more responsive both to consumer expectations of the quality of farm produce and to town-dwellers who want to revive their relationship with the countryside.

If the rural and farming population can't tap this new dynamic, no new farms will be set up, and the system will collapse. Financial prosperity can flourish from within healthy rural communities. This was our experience in the Larzac: in a poor area, we created a rich community linked to struggle. People settled there not knowing how they would manage financially, but wanting to live collectively. Together, we created the conditions for working the land, established common access to tools, set up outlets for direct sales of meat, and started a cheese co-operative.

FD: Young people from sectors other than farming will be the saving grace in this agricultural renewal. But success depends on goodwill from corporate agriculture and a more general opening up to fresh ideas.

SUSTAINABLE FARMING

You've moved from basic trade-union activity to the fight against the entire world of intensive agriculture. How fundamental is this break going to be?

JB: When you go beyond the defence of working conditions, incomes or jobs to challenge the social and ecological purpose of work and human activity, then I think that's a complete break.

FD: This break is best expressed in what we refer to as 'sustainable farming'. It's still new, but it's a project which we have developed progressively through our struggles against intensive farming. It's the fruit of our thinking over the last twenty years, reinforced daily by work on our farms. It's a coherent system which simultaneously integrates different farming techniques and values.

'Sustainable farming', for us, is like a flower with many petals. Everything holds together: the status of farmers, the income, work-sharing, the quality of produce, the ownership of farms, respect for natural resources, equality between North and South. All these elements, like the petals on the flower, are inseparable, and if one is missing, there is an obvious gap. We're not talking about a 'model'; it's an outlook, a different conceptual approach to the job. Some farmers are organic, some are not; some are cereal growers or pig breeders, but all are working towards the same goals.

JB: Certainly, compared to the model imposed on us all by the FNSEA, it's a radical break. As François and Françoise have shown, it is possible to reject the intensive model of farming, and in fact many people have done so successfully. The need to find alternative ways of farming dates back to 1980, when our movement publicly denounced the use of growth hormones in calf breeding.

Our comrades became aware of the economic and ecological madness of a system which consists of separating the calf from its mother, only to give it milk which has been collected by lorry, taken to the factory, pasteurized, creamed off, dried, reconstituted, packed, and then returned to the specialist calf-breeder to feed the calves. Heavy subsidies from Europe to the dairy industries ensured that the reconstituted milk was less expensive than the natural stuff. Such crazy methods led us to examine the real purpose of our work, and prompted us, in 1992, to develop the concept of 'sustainable farming'.

Does sustainable farming meet general social needs?

FD: Sustainable farming links choices concerning the means of production with local solidarity and the protection of biodiversity. Sustainable farming is defined by an 'approach' and a 'perimeter'. The 'approach' is the direction, the horizon to be aimed at, regardless of the situation on one's farm. Ten principles define sustainable farming.[1] They deal with how animals are fed, illnesses are treated, plants are protected; and how a balance is created between capital and work.

The 'perimeter' defines the extent of farmers' activity, sets limits to intensive farming – the number of animals or the quantity of nitrate per hectare, for instance. It lays down guidelines for the size of workshops, and so on. This policy does not turn its back on those who have remained prisoners of intensive farming. However, in insisting on moving away from this system, we emphasize that changing the method of development is pointless if the social dimension is not taken into account. The Farmers' Confederation, especially in the west of the country, is heavily involved in defending small and medium-sized pig and poultry farms threatened by bankruptcy as a result of the crisis in those sectors. We have to struggle to master all the various skills involved: how much feed to give, how to cost work, how to ensure that contracts are fulfilled. Many breeders have rejected intensive farming, but don't have the practical or legal means to opt out. I know what I'm talking about, as it took us several years on our farm. One method of farming can't simply be substituted for another overnight.

JB: This is where we come back to the roots of trade unionism and the example of the Jura Federation's alternative to Marxism. When the First International was founded, two opposing currents surfaced – on the one side Marx, and on the other Bakunin – with two different ways of organizing the workers' movement. Marx's interpretation of trade unionism centred mainly on the recuperation of surplus-value, and the worker's position in the face of capitalism. But I'm very attached to the history of the Jura Federation, which shows Bakunin's approach in action.

The watchmakers' union in the Jura grouped together the farm labourers and other workers. They were organized in small workshops. Each person

1. See Appendix 2.

was autonomous and responsible for their own work. The Jura experience found an echo when the Lip workers, after a long strike at the beginning of the 1970s, took over the production and sale of watches under self-management.

Another example is that of the building workers in Spain, organized by the CNT (a Spanish trade-union confederation), who refused to build prisons in the 1920s. Not a single prison was built in Spain between 1920 and 1930; the building workers chose unemployment rather than construct such buildings. In 1936, right in the midst of the Civil War, the union movement in Spain organized self-management in the factories and the collective running of the land. These are the sorts of examples that we in the Farmers' Confederation have drawn on, placing social considerations at the forefront when it comes to deciding how work should be organized.

The sustainable farming approach includes an emphasis on improving the quality of food. Have you considered introducing labels on produce to highlight this for consumers?

FD: At the moment it's not on our agenda, although your question does touch on a real concern for consumers, who want to know what they're eating. We would like to see the introduction of labels that identify the origin, ingredients and method of production: for instance, whether chickens and calves have been reared outdoors and fed cereals grown on the farm, or indoors and fed with feed-industry-produced fodder.

JB: The *Appellation d'origine contrôlée* (AOC) system, familiar to many people because of its use in wine labelling, points a way forward here. Initially, the AOC label, backed by government legislation, was a means of protecting a product from being copied. Today, it has evolved into a label that details not only the product's identity but also the conditions of production. It's a big change, and it has provoked widespread debate at the base of the movement. If you add together the various initiatives – AOC, organic farming, changes in farm culture, sustainable farming – you begin to get a strong feel of a new farmers' movement which, I believe, will eventually marginalize industrial agriculture.

THE COUNTRYSIDE: A MAN'S WORLD?

How does this new approach to farming deal with the issue of women in agriculture?

FD: That's a sore point. I remember, at the time of the CNSTP, there was a very active women's group which didn't limit itself to the issues of maternity leave or pension rights for wives. This group discussed the status of women, their role on the farm, job-sharing and male chauvinism. But their activity didn't have much of an impact inside the Farmers' Confederation. That was a great shame, because it prevented us from thoroughly reorganizing rural life.

Today, the profession seems to be increasingly male-orientated; the overwhelming number of farms are being taken over by men, while women follow careers outside farming. As we said before, many farmers remain unmarried. There's no reason to see the countryside as essentially male, but such an imbalance makes any demand for equality futile. A new women's group has recently been set up in the Confederation which, we hope, will address these problems.

Surely, changing the form of agriculture involves making more room for women in farming and developing equal opportunities?

FD: I'm not sure that the problem is between men and women. I see the issue more as an acknowledgement of status for all farmers, both men and women. Defining rights and responsibilities for all is not a male–female question.

JB: I must admit, however, that the intensive form of farming has marginalized women still further. Farming became increasingly technical and this, together with a macho drive to produce ever greater quantities, reinforced women's exclusion from the farm. Women are kept out of farming in other ways, too. For instance, loans are often more easily obtained from the banks if the wife works outside the sector.

Can sustainable farming contribute towards the feminization of the job?

JB: Yes, I think it can. Organizing the work differently could contribute to achieving equality, but a change in the whole conceptualization of agriculture is also required: farming must cease to be competitive and develop a different relationship towards animals and the soil.

José, are there women in your GAEC?

JB: At this moment there's one, Danièle Domeyne, and a second is applying to join. The diversity of our production allows for choice in the work undertaken and the hours worked. Danièle, who is an agronomic engineer, is in charge of cheese production. The new candidate is interested in looking after sheep – that's what I do at present – as well as crops.

And you, François – how does the division of labour work on your farm?

FD: On the farm, Françoise is responsible for the milking. We have a mechanized system, so it takes only forty minutes, mornings and evenings. She does it during the week, and I usually do it at weekends. Before I took on national responsibilities in the union, we would share the work: I would milk in the morning and Françoise in the evening. Apart from that, she takes charge of the farm visitors, and her public-relations skills have ensured that camping on the farm has become a huge success.

Then there are the children, who are also part of the farm. All five have shown a lively interest in what goes on. My daughter loves horses; the boys are more interested in the orchard, looking after the cows and the farm equipment.

8

FREEDOM FROM

'FREE TRADE'

A WORLDWIDE DICTATORSHIP

The fall of the Berlin Wall coincided with an acceleration of free trade, leading in turn to an increasing concentration of big industry. Agriculture was not spared in this race. The arrival of biotechnology further hastened the process. The future of the world food market rests with a handful of agrochemical firms. How do you react to all this?

FD: I prefer to talk about globalization rather than world markets; the word has passed into the language as a symbol of all the evil caused by the unbridled liberalization of trade. Globalization is treating the planet as one vast commercial domain, where no rules or restrictions apply, and goods are exchanged with no heed for social, ethical or environmental values. It's the hegemonic market, intent on devouring everything.

JB: International relations reflect the technology of a particular time and place. Thus, under the Roman Empire, the centre of the world was around the Mediterranean basin. As new continents were discovered, this centre grew progressively until it encompassed the whole globe, creating a new concept of what the world was – one where the self-proclaimed, colonial 'centre' appropriated land and trade. This then defined the relationship that Europe, land of explorers, entered into with the Americas, the Caribbean, Africa, Oceania, and to a certain extent Asia. Means of transport and communications ensure that today's market is genuinely worldwide. As far as world leaders are concerned, the entire planet should submit to market laws. Our struggle is based on resistance to this development.

There are two different views of society. One where the market, with its own rules, runs everything, and where all human activity (health, education, culture, and so on) takes place with capital as the bottom line; the other where people and their political institutions – not to mention issues such as the environment and culture – are at the forefront of society's concerns.

FD: Globalization is also standardization from below, deregulation through progressive erosion of all fundamental rights, trade with complete disrespect for the needs of people. When you consider, for example, the political conditions of the shady lobbies who imposed the AMI, it's clear that dominance of capital motivates those who want to make a profit out of everything.

JB: It's a worldwide dictatorship; if you're not in the market sphere, you're a nobody. We no longer live under conditions of traditional management and interstate conflicts, but in the middle of a war between private powers, with the market as the battleground. To understand the extent of this, all you have to do is look how the traffic in money makes more profit than traditional production and trading activities combined. Today, money works by itself. This has produced a new species of parasite: vampires thirsty for money. Money addicts.

So what you object to in globalization is the traffic in money rather than the exchange of goods?

JB: We reject the global model dictated by the multinationals. Let's go back to agriculture: less than 5 per cent of agricultural production goes on to the world market. Yet those responsible for that 5 per cent of international trade dominate the other 95 per cent of the production that is destined for national consumption (or neighbouring countries), and force this sector to submit to their logic. It's a totalitarian exercise.

THE WTO'S LIBERAL PROJECT

Where does Europe stand in relation to this free-market exchange between the USA and the Cairns[1] group of countries?

JB: We need to go back a little over the history of Europe and GATT. In 1957, the CAP was created in pursuit of food self-sufficiency. As we've seen, it represented a unique market between member states based on a range of single products, and offering guaranteed prices – good or bad, depending on the product. For cereal and beetroot growers these prices were very high; for dairy and beef producers less so; and for all other products minimal or even nonexistent. This common market was closely guarded by the Community: imports were taxed by amounts which varied according to the difference between the price guaranteed inside the country and the fluctuating world price. For export surpluses, the same system of variable subsidies was applied to industrialists and exporters to compensate for the price differentials between the internal and external markets.

1. This group, founded in Cairns, Australia, in 1986, brings together fourteen agricultural and textile exporting countries, including Argentina, Australia, Canada and New Zealand, which are in favour of liberalizing world trade.

The system works because it receives financial backing from the EU member states; each country contributes to the European budget according to its GNP and its VAT receipts. Thus assured of their prices and sales outlets, European agriculture and the agricultural feed industry (which have both been pushed into intensive production) strove to export a surplus in order to increase their share of the world market. This led to an explosion of the European budget. It should be pointed out that they were encouraged in this by governments – especially in France, where the slogan was 'Produce to export: agriculture is France's green petrol.'

Instead of establishing policies to control production, Europe, influenced by the agribusiness lobby, did not intervene. However, as a result of widespread criticism and budgetary stalemates, the EU was forced to reform the CAP in 1992 by lowering prices in order to attract a bigger share of the world market. This reform represented a first step in 'total liberalization': a partial but significant dismantling of the system of variable taxes, which were replaced by fixed Customs duties. Given that European agriculture – even the supposedly lucrative cereal crop from the Paris basin – is unable to compete on a world scale, direct aid is given to producers by the EU. For free-market liberals, this reform was only a first step in a campaign to conform to GATT's policies.

FD: GATT was the precursor of the WTO. GATT is too loosely structured to be called an organization; it's more an agreement which other countries can enter on a voluntary basis. Until 1986, and the 'Uruguay Round', agriculture and food were not the remit of GATT: each country or group of countries was free to adopt the agricultural policy of its choice. But GATT, greedy for more, appropriated control over agriculture as well.

The philosophy and principles of GATT are clear: they are those of a free market. Customs duties must be lowered – the state must treat imported products in the same way as the national equivalent, and abolish any preferential agreement between states – the clause known as 'favoured nation status'.[2]

Take the example of bananas, an issue which pits Europe against the USA: seventy countries from Africa, the Caribbean and the Pacific (called APC countries) have links with the EU through the 'Lomé Convention'. Certain products from Martinique, Guadaloupe and the Canaries are bought by the EU at an inflated price compared to their value on the world market to help these APC countries and protect their agriculture; for instance, they are guaranteed an import quota of 857,000 tons of bananas for the European market. This agreement is not very popular with the American multinationals, which have their own plantations in South America, producing bananas more cheaply. As a result, in April 1999 the USA obtained a condemnation of Europe's policy from the ORD, a WTO body which settles disputes. According to this philosophy, the free market is supposed to stimulate economic growth and contribute to general prosperity. One cannot, however, equate growth with development, as was made clear by CNUCED, which has denounced the growth of inequality in the world since the establishment of the WTO in 1995.

The WTO has extended its sphere of intervention beyond the regulation of trade to the imposition of the free-market model, requiring some countries to deregulate. A state cannot refuse to import agricultural or food produce on the grounds of protecting the health of its population and livestock. Such protective measures have to be accompanied by scientific arguments, backed by international experts recognized by GATT.

This was the *raison d'être* of the 'Codex Alimentarius': to establish health norms. Today, the national delegations of this body have been infiltrated by food industry representatives who can dictate its rules. The EU and USA between them send 60 per cent of the delegates, but represent only 15 per

---

2. According to which any preferential treatment accorded to one country must automatically be extended to all others.

cent of the world population! In 1997, it was the 'Codex Alimentarius' that proposed the international ban on the trade in products based on untreated milk.

JB: The reform of the CAP in 1992 meant a lowering of community protection, resulting in a drop in farmers' income which – so free-market liberals maintained – would encourage them to be more competitive. To soften the blow, Europe implemented a system of direct aid which, in principle, was not linked to levels of production. Rather, the yardstick used was the area of land cultivated, or the number of cattle in the herd. For cereal- and oil-producing growers, no upper limit to the aid was set, and prices were dependent on local yield, so that the richest areas, where land was irrigated, would get more compensation. In other words, it was decided to give help to those who were already wealthy, thereby widening rather than reducing the income gap between farmers.

The 'cereal bonus' was worth more than 3,000 francs per hectare in some departments of the Paris basin, but only 2,000 in Bourgogne, where the yield was, on average, smaller. For maize that has been irrigated, there can be a difference of between 600 and 1,000 francs, depending on the Department. This was a real scandal, endorsed by the EU and imposed on the French government by the FNSEA. This situation raised issues about the size and purpose of the public funding of agriculture (70 billion francs of direct aid annually), 80 per cent of which goes to just 20 per cent of farmers.

The reduction in community protection hit hardest in those sectors which were already suffering, such as fruit and vegetable farming – already extensively relocated to North African countries – or mutton rearing.

FD: In March 1999, in Berlin, the EU proposed 'deepening the reform of the CAP' as a liberalization measure to prepare for the imminent Millennium Round of the WTO discussions due to take place in Seattle. They stressed that the lowering of prices by 10–20 per cent as a result would be partially compensated by direct aid.

GLOBALIZED TRADE, GLOBAL DEMANDS

In November 1999, Europe went to the WTO meeting in Seattle, to negotiate the next round of liberalizations. You went too. Why?

FD: Those negotiations were crucial, particularly for people concerned with agriculture and the exploitation of forests. At stake was the fate of farmers worldwide. We went to Seattle because we had had the Marrakesh agreements in April 1994, when the WTO was set up; the agreements made then were scheduled to run for seven years; they obliged every country to reduce its Customs protection by 36 per cent and to allow imports of a minimum of 5 per cent of the volume of its internal market at reduced tariffs. This liberalization hugely benefited agribusiness. The reduction of state-subsidized exports and Customs duties meant a significant drop in support for internal market prices.

JB: The Marrakesh agreement envisaged drawing up a balance sheet of prosperity around the world prior to the new round of discussions at Seattle. We're still waiting for that. But we don't need an official report to know the extent of the damage caused. Between 1992, the first year of the new CAP orientation towards the world market, and 1998 (halfway through the timespan of the Marrakesh agreements), Europe had a million fewer active farmers. In France, 300,000 were lost.

As for the countries of the South, they have been the biggest losers in these negotiations, because they can't afford to give direct subsidies to their farmers. Opening up their frontiers is a direct attack on their subsistence agriculture, and exacerbates the exodus from these countries. For example, South Korea and the Philippines, two countries that were self-sufficient in rice production, are now compelled to import low-grade rice at a lower cost than the local crop. This creates an imbalance in the national market.

Similarly, India and Pakistan, big textile and cotton producers, are now forced to import industrial fibres that compete with their own products. It's the same in Europe. The quota policy of milk production established in 1984 is being challenged by the rule that at least 5 per cent of internal consumption must be imported. This extra 5 per cent increases the price, and is a burden on the European cattle-breeders, who have suffered in the last three years.

That's why the Farmers' Confederation, along with some thousand organizations from over a hundred countries, were demanding that before any new trade deals were agreed by the WTO, an assessment of the economic, social and environmental consequences of the original agreement in Marrakesh should be undertaken.

So you went to Seattle to say 'No to the WTO. . . .'

FD: Not only that. We wanted to follow the world negotiations closely so that we could anticipate their repercussions for farmers and, more particularly, for the CAP. The Farmers' Confederation appointed a delegation of four (Patrice Vidieu, general secretary; Christian Boisgontier, our delegate on the European Farmers' Co-ordination, José and myself), and the French government arranged accreditation for us at the WTO. We took along proposals for discussion with other farmers' organizations in the *Via campesina*.

Could you sum up your proposals?

FD: In the first place, we reaffirmed people's rights to control what they eat and to choose their form of agriculture freely and democratically. The abundance of goods and food has reached unprecedented levels, but so have the number of homeless, unemployed, or undernourished people. Europe based its agricultural policy on the concept of food self-sufficiency by protecting its markets from external competition. We believe that this was legitimate, and that all other countries – or groups of countries – in the world should be allowed to choose which crops they want to protect from competition to ensure self-sufficiency in food, and maintain a balance between town and countryside.

This has to include having the power to oppose the relocation of agricultural production that European agribusiness is currently undertaking – setting up pig and chicken farms, and greenhouses for the cultivation of vegetables, in countries where costs are lower and labour and environmental regulations are barely enforced. In the majority of instances, these practices have distorted local agriculture, destroyed the environment, reduced natural resources and threatened food safety. When large agricultural enterprises replace a thousand farmers, while simultaneously increasing production and selling it on the world market, national food policies are jeopardized. Thus in Brazil, a major exporter of agricultural produce, we see a growing proportion of the population suffering from malnutrition. Many families have been denied access to land, and thus the possibility of farming for themselves, and the result is widespread hunger. Each country – or group of countries – is entitled to protect its food resources. This is the basic tenet of food sovereignty; it means that protection from imports is an indispensable condition for international fair trade.

JB: That was our second principle: all international exchanges should be governed by the principle of fair trade. Fair trade means buying goods at their real production cost, instead of the situation today, where world prices are the result of 'dumping'; low prices are set by the wealthiest countries using big handouts to cut export costs and other disguised internal financial help. Powerful companies frequently adopt such practices in order to undermine the prices in the countries they intend to move into next. Once they've swept the local agriculture out of the way, they raise prices again. So, for example, the export of European frozen meat, which is very highly subsidized, has resulted in halving Sub-Saharan livestock. That's why we're calling for the abolition of all export aid, plain and simple.

The effects of the Marrakesh agreements prove that liberalization of agricultural trade has not led to a more stable world market, nor has it improved the food supply for the poorest countries. Fair trade implies a ban on all forms of 'dumping'.

FD: The monopoly – or quasi-monopoly – whereby a handful of companies share out the world between them is incompatible with the principle of fair trade. Take the example of water. The world's water resources are in the hands of a few transnational companies.

In France, Vivendi has taken control of the treatment and management of water and waste, as well as of communications and the running of hospitals. It is increasingly extending its empire in the Third World. By mergers and takeovers, a few multinationals have accumulated economic power well in excess of that of many countries. General Motors' turnover is bigger than Thailand's GDP. In the same way, the multinational Cargill, after merging with Continental, now owns 40 per cent of the world's maize production, a third of the soya crop and 20 per cent of wheat cultivation; moreover, its alliance with Monsanto means that it has control through the whole food chain, from seed to plate.

These monopolies are operating unfairly, because they exacerbate inequality. The history of the last twenty years is littered with thousands of examples. The OECD countries, representing just 19 per cent of the world population, monopolize 71 per cent of world trade;[3] the 225 wealthiest people in the world have at their combined disposal the equivalent of the annual income of 47 per cent of its poorest inhabitants – 2.5 billion people.

In calling for the abolition of export aid, aren't you in agreement with the WTO and the Americans . . .?

FD: Yes. In calling for an end to all export subsidies we concur with the WTO. But that's as far as it goes. For, as José has explained, this abolition

3. UNDP Report, 1999.

must go hand in hand with a right for countries to establish Customs barriers according to their needs. Developing countries can impose higher tariffs. Of course, the industrialized countries are free to enter agreements with Third World countries, as Europe did with the ACP countries.

In this way, the world market will reflect the real cost of production for the exporting countries. From such a basis, fair trading can begin. Far from being protectionist, this is the only policy compatible with sustainable development. The practice of pitting farmer against farmer must stop.

JB: Some countries have a choice in what they export; others – because of geographical or climatic constraints, or the density of their populations – depend on food supplies from abroad. This means that the same rules governing agricultural trade can't be applied to all countries.

FD: The EU exports basic foodstuffs such as cereals, white meat, milk powder and beef, and would like to export more. But the market is already saturated with the surplus from large producing countries such as the EU, Canada or the United States, or from countries whose agricultural models are based on the ranch system, including Australia, New Zealand and Latin America. We don't think Europe has the space to produce these basic items for export, but if producers want to do so, they should not be given subsidies. It's completely irresponsible – not to say costly for the taxpayer – to drag European agriculture into the quest for an ever larger world market share. We should show solidarity with other farmers around the world by not helping to create imbalances in their local markets.

JB: On the other hand, the EU has the potential to be a leading exporter in a number of fields requiring specialist knowledge of specific produce, such as wines, spirits, cheeses, mustard, foie gras, and so on. These foods are usually produced in specific geographical areas, and are often linked to a local culture. They aren't subsidized, and have acquired a market as a result of their quality and price. There's no reason for this to change. Such production does not stop Third World countries from developing their own agriculture.

GETTING TOGETHER IN AMERICA

You left eight days before the start of the WTO Summit and went on a tour of the States, including a visit to Washington. Why, and what happened?

JB: Twelve of us went – four official representatives, along with delegations of militants from the Aveyron and from Brittany. We wanted to visit American farmers and ordinary citizens who had expressed their solidarity during our spell in prison: our friends from the NFFC (National Family Farm Coalition) – a federation of thirty-six family agricultural unions, organizing some 90,000 families; ecologists from Friends of the Earth; Public Citizen, the most important American consumer association, set up by Ralph Nader; and the Institute for Agriculture and Trade Policy (ATP).

FD: The NFFC invited us to take part in the first press conference of a farmers' organization against GMOs. Along with Friends of the Earth, we demonstrated in Washington and then visited some farms. In Pennsylvania we met farmers who were under threat from low prices and concentration of production.

When, on the other hand, we travelled to the Seattle area, we witnessed the success of organic vegetable growers who sell their produce through local consumer networks. In the town centre, José, speaking in English, addressed a crowd of 2,500 people attending an international forum; after describing our campaign, he reminded the audience of the founding act of the United States, when, in 1773, the inhabitants of Boston threw a cargo of tea into the sea to protest against unfair trade.

The media coverage of our visit helped to make our ideas known in America, and allowed us to explain that our actions had not been prompted by French or European chauvinism, but had a global relevance.

So how was your visit there?

JB: We went to Seattle to join the demonstrations aimed at stopping the WTO Summit. The city was crawling with people and organizations from all over the world, as well as from the USA. It was an excellent opportunity for a wide range of informal meetings with other militants, and we joined together in human chains, sit-ins, demonstrations, tree-plantings, debates, and so on. It was the opening shot of a co-ordinated worldwide response.

FD: The French delegation organized through the CCC-OMC (Committee for Citizens' Control of the WTO) would meet every day at the Speakeasy, a cybercafé in Seattle, to take stock of what was happening. Dominique Voynet was the only Minister who came to see us because, as she put it, 'we knew more about what was going on than she did'. MPs Jean-Claude Lefort, Guy Hascouët and Harlem Désir, together with Senator Jack Ralite, also spoke with us, but no other member of the French government delegation saw the need to come and talk with the CCC-OMC.

The issues on the agenda were not just to do with agriculture. Besides the question of food, which concerns everyone, issues such as public services, cultural production, intellectual property rights, and finance were involved. There was a widespread coming together against a world being run without transparency or democratic rule. *Via campesina* used the Seattle conference to organize a big workshop over several days on the issue of globalization, attended by seventy delegates from thirty countries.

Not all opponents to the WTO take the same position. Some oppose any world trade organization; others, like you, call for its transformation. Could you describe these differing viewpoints?

FD: Two clear positions emerged within *Via campesina*: on the one hand, there are those who hold that 'the WTO is nothing to do with us', and believe that agriculture and food issues should be withdrawn from the negotiations. Some of these people maintain that the EU is no better than the USA, and the hardliners say 'Down with the WTO'. Then there are those, like ourselves, who consider that a world regulatory body for trade is needed, but that its rules and the way it functions must be transformed.

We didn't expect one side to convince the other. In any case, these positions aren't so different as they may seem, because they're united in their assessment of the harm done by the WTO. You can't talk about factions within *Via campesina*, which is a worldwide farmers' organization defending what it considers to be the most crucial issues of the day. What holds for Santiago or Bamako doesn't necessarily hold for Rome or Paris. The exchange of opinions and experiences makes this a wonderful network for training and debate. The delegations to the *Via campesina* don't negotiate in terms of conquering the market but to promote, above all, development of mutual respect. This 'farmers' Internationale' represents a living example of a new relationship between the Northern and Southern states.

Fundamentally, we're all agreed on our view of agricultural and food issues: respect for the right to food sovereignty, in favour of sustainable agriculture and access to land through agrarian reform; opposition to trading in living organisms, which results in GMOs, agricultural patenting, and bans on farmers being allowed to sow their own seeds.

JB: In discussions with the Americans, I also noted the differences that have been mentioned: some workers asked why we wanted rules while simultaneously calling for food sovereignty. We exchanged ideas, while maintaining respect for the views of those who differed with us.

I spent a long time explaining our idea of submitting disputes to an international trade tribunal independent of the WTO. The WTO is, above all, a political organization set up by the affiliated governments and countries. It is a legitimate world organization, but it has very quickly turned into an exclusive tool of commerce.

THE STALEMATE OF CHAUVINISM

Some of those experiencing the destabilizing effects of globalization seem ready to retreat into nationalism. How do you differ from these chauvinists?

JB: Their idea of sovereignty relates to the nation-state, and theirs is a selfish, frightened and irrational response. Since the fall of the Berlin Wall, the pace of world trade has accelerated, and needs to be regulated. Our concept of sovereignty enables people to think for themselves, without any imposed model for agriculture or society, and to live in solidarity with each other. This sovereignty means independent access to food: to be self-sufficient and to be able to choose what we eat.

Furthermore, the globalization of trade must be counteracted on the same level – that is to say, on a world scale rather than on a narrow-minded nation-state basis. Nationalists worry about the mixing of races, whereas we welcome fair trade, cultural exchange and solidarity: we stand for a dignified and free life under real democracy. Nationalists don't want to hear this; they prefer to make scapegoats of other countries. Since 1957, French agricultural policy has been formulated in a European context, and there would be little point in defending our farming by itself. It's not a case of protecting French farmers but of defending a model of development. France could have used its weight in Europe to do this, but our elected representatives are entrenched in old models of intensive agriculture.

The French Prime Minister, Lionel Jospin, falls between two stools when he tries simultaneously to promote sustainable farming and France as a major agricultural exporter. Nationalists want strong borders so that the French multinationals, especially those concerned with food, are masters at home and bosses everywhere else – on our money, of course.

It's dangerous and illusory to take pride in the fact that France is the leading exporter of farm produce in Europe, when 90 per cent of our exports are subsidized.

Nevertheless, expanding exports is one of the FNSEA's warhorses. How do you counter this line?

FD: The FNSEA was present in Seattle, working as a pressure group in the corridors of the official Conference. They refused to join us on the street, in the demonstrations. Their position was to follow the French government and the EU, as they usually do, and to defend the European intensive model.

JB: Basically, the politics of the FNSEA are those of 'every man for himself'; they want protection of the internal market and public subsidies for their exports. They want to have their cake and eat it.

THE WORLD IS NOT FOR SALE!

Let's go back to Seattle. Hundreds of thousands of people demonstrated, paralysing the town centre, preventing delegations from leaving their hotels, and causing the opening of the Conference to be postponed. The delegates realized that an agreement would not be reached, and there's little doubt that the street protests played a big part in this. How did you see the day?

FD: For me, the first sight of the demonstration was full of symbolic meaning. Five *Via campesina* farmers, including Rafael Alegria, the general secretary, and José wearing the green cap of the farmers' movement, were at the head of the crowd, hand in hand with members of the AFL–CIO, the largest American trade union. It was an important signal: that in the first mass demonstration of trade unionists and ecologists, farmers were at the front. It's a particularly powerful image for Third World countries, where the majority of the population are farmers or live in rural areas. I recalled with emotion Bernard Lambert's words when he declared, on the Larzac: 'Never again will farmers be like the people of Versailles' (who opposed the Paris Commune in 1871).

JB: The good news about the demonstration was the emergence of a young, radical movement that brought together dozens of groups in the USA under the banner of the Direct Action Network (a non-violent radical movement). It was a sight to behold: young people blocking the crossroads beneath the Sheraton Hotel, determined but non-violent, quiet, friendly . . . and unbelievably efficient in paralysing the town. And the not-so-young, women and men, holding hands with the young people to block the hotel exits. A folk group encouraged people to dance, while the WTO delegates were trapped inside. Behind the hotel windows were the officials, imprisoned in their virtual world while below, the real world was rejecting them. This time they couldn't ignore what was happening, because the world's media were present. Grenades, plastic bullets, tear gas – the National Guard had to be called in to remove the non-violent demonstrators. I have distinct memories of all sorts of wonderful things: grandmothers changing the words of traditional songs into protest songs; an American steel band, chanting in French 'Tous ensemble, tous ensemble, tous ensemble' (All together . . .); the courage of a group of lesbians, naked to the waist in the chilly November weather. I had the feeling that a new period of protest was about to begin in America – a new beginning for politics, after the failures and inactivity of the previous generation.

How do you account for the success of the Seattle mobilization?

JB: The mobilization against the AMI in 1998 was a kind of springboard that launched a new network of like-minded activists. The McDonald's action in Millau further strengthened the movement; people began to realize that globalization affected their daily life, rather than being something remote and abstract. These days, in our post-industrial society, social awareness against alienation is more likely to come from thinking things through than from experience of more traditional overt exploitation. This is not so under a dictatorship, or in the neo-colonial economies, where suffering is a constant reminder of who the enemy is. In Millau, we succeeded in making something abstract become relevant. In Seattle, we found the echo of our actions in Millau.

FD: What occurred was a convergence of the movements in the countries of the North – unions, ecologists, consumers, civil and gay rights activists, to name but a few – with the countries of the South.

Links were made that have not been seen in the USA for a very long time. Some American colleagues admitted that they hadn't seen anything like Seattle since 1968, when the anti-Vietnam War demonstrators in Chicago stopped the Democratic convention.

We took advantage of the divisions in the WTO: divisions between the rich countries squabbling over market shares, Bill Clinton prevaricating over what outcome would better further the interests of the Democratic Party in the November 2000 presidential elections; finally, there was the dissent at the treatment of the Third World countries which led to seventy-five countries rejecting the agenda proposed by the big four: the USA, Canada, Europe and Japan.

The demonstrators won. But what has been gained? The WTO continues to exist, and there are even rumours that the USA has obtained the right to trade without regulations.

JB: Sure – despite Seattle, free-market liberalism is still rampant, the WTO is still alive and well. You can't put an end to either with one demonstration. Our objective was to stop the extension of the WTO's powers. In this respect, Seattle was a real, not just a symbolic, victory.

FD: From now on, no similar international negotiations will be able to dodge the questions of transparency, fair trade or democracy. WTO boss, Mike Moore, and Pascal Lamy, its European delegate, were forced to recognize this. I believe that we made ourselves clearly understood.

Agriculture should not be reduced to mere trade. People have the right to be able to feed themselves, and take precautionary measures on food as they see fit.

JB: Health, education, culture, food – these are all issues close to everyone's heart. Today they are in danger of becoming mere commodities. For example, medicines that could eradicate certain endemic diseases in Africa are not supplied because the local economies can't pay for such proprietary drugs. Waves of opposition to this commodification can be felt from all corners of the world, calling for a return to the primacy of politics over capitalist economics.

Do the Seattle gatherings represent a new internationalism?

JB: There are no preconceived ideas. Those days have gone – thank goodness – when popular movements were slotted into theoretical constructs. Seattle showed the opposite. People came together not with any worked-out theory, but to take action. From wherever they came they brought their experiences and points of view, often finding common ground. For far too long, theories and analyses have been shuffled around, promising change. People today have lost confidence in these theories. Seattle revealed the existence of an informal worldwide network. There were no red flags in Seattle, no portraits of Che, no ideas of socialism in one country; that's all finished.

Fortunately, issues such as health, food, the environment, water resources, can be fought for, and progress can be made without the old theories and forms of organizing, as *Via campesina* shows. Who would have thought, thirty years ago, that there could be an international farmers' movement without ideological tendencies? Who would have imagined that dictators such as Pinochet or Hissène Habré would have to account for their crimes because of social pressure and defence of human rights? Agriculture has become a major issue. It's the activity all countries have in common, and it's rapidly becoming the central focus for protest and resistance.

FROM SEATTLE TO BRUSSELS

As far as the EU is concerned, how significant was the defeat of the WTO in Seattle?

FD: Things are different after Seattle. We're reconsidering the Berlin agreement on a Common Agricultural Policy, which had previously been envisaged within a context of further liberalization of world trade. The victory at Seattle means that we should now clear the decks for a new CAP which reorientates agriculture to meet the needs of the consumer.

Can you be more specific about this new CAP that the Farmers' Confederation is proposing?

FD: There are four main objectives. The first is to satisfy the European consumer in terms of quantity and quality of food; the second is to promote a form of agriculture suitable for large numbers of farmers; the third is to achieve a level of income for farmers that is in line with other sectors of the economy. Finally, we want a fair world trade system based on co-operation with the less developed countries.

These objectives can be achieved through the reintroduction of community protection, with a system of import tariffs that can be varied according to world prices. Once back in place, such a scheme would allow an internal market to function, where prices of agricultural goods are determined by the efficiency of different regions. Prices may need to be topped up to compensate for geographical disadvantage or natural disasters, but direct subsidies would be small, and justified by particular social and environmental considerations. Receiving the correct price for their produce would enable farmers to concentrate on satisfying consumers.

At the same time, direct and indirect support for exports must be removed as a precondition for building international co-operation, especially with Third World countries.

JB: The new CAP should also deal with two further issues. First, countries should be allowed democratically to set their own safety criteria. These would probably become more stringent than current international controls. Secondly, a new CAP must champion each nation's ownership of its genetic resources against the biotechnology industry's attempts at 'theft by patent'.

Some of the stages of this economic reorientation can already be sketched out – pending fuller consultation in due course. In the cereal sector, apart from the removal of direct subsidies, it will be necessary to withdraw all funding for wealthy regions while maintaining it for underprivileged areas.

The decision to end milk quotas by 2008 should be revoked. Instead, the policy of volume control should be extended to include other agricultural produce, and a new redistributive policy should be adopted to help cattle-breeders. In 1984 there were 470,000 milk producers in France; today, a mere 130,000 remain, and if the current CAP continues, only 75,000 will be left by 2010.

DEMOCRATIC CONTROL

What are your plans for building on the success of Seattle? Do you have alternative proposals on world trade?

JB: The regulation of international trade is a good thing, so long as it's based on equality of rights, not on the dominance of the economically strong. As our banner in the streets of Seattle proclaimed, we are in favour of the WTO adopting the Human Rights Charter. Why should the global market escape the rule of international law or human rights conventions passed by the United Nations?

The WTO has arrogated the functions of legislature, executive and judiciary solely for itself. In the eighteenth century, such an anti-democratic concentration of power provoked the French Revolution!

To break this monopoly of power, we have demanded an international court of justice, composed of professional lawyers, independent of the WTO, whose rules are based on the Declaration of Human Rights and other conventions agreed by the UN. This court could hear appeals by countries dissatisfied with WTO decisions. This arrogation of all legal power by the WTO contributed, no doubt, to the collapse of the negotiations in Seattle. Political consensus must be built from the local to the international, from citizens to the state, not the other way round.

The WTO isn't going to change overnight. We're in for a long struggle. Building on the international gains won at Seattle, we're working towards setting up a permanent watchdog[4] in Geneva, seat of the WTO. This centre will provide information for all those mobilizing on the issue of world trade. We also intend to provide training to enable activists to find their way around in the jungle of international institutions: WTO, IMF, World Bank, et cetera.

We want to ensure that the WTO knows it is constantly under scrutiny. I call this the 'Dracula principle', something we started with the struggle against the AMI: Dracula, the vampire, can't bear the light; nor could the AMI. We want to open all the windows on the WTO as well.

The networks should organize on an international level, but we don't want everything centralized. Movements must keep their autonomy; we have to organize our own schedule of activities. If I may make another comparison with the French Revolution: in 1789, the Estates General developed village by village, in harmony with one another; it was the sum of this process of alliances that finally tipped the scales. I believe we should stick to this approach, and find cohesion in the demands that arise from the coming together of movements. The blockade of the Sheraton Hotel in

4. This project is known as Global Citizen Initiative.

Seattle was like the storming of the Bastille; it will take a long time, but we must move forward to the Constituent Assembly.

FD: Among the first tasks of our watchdog organization could be a thorough examination of the Marrakesh agreements; how to improve the transparency of the WTO, the role of the ORD; and how to promote an international court of appeal outside the WTO.

But your watchdog project will have no effect on the WTO delegates' lack of legitimacy, nor on citizen representation at a world level. . . .

FD: That's true. Charlene Barshefsky, leader of the American delegation, didn't even have a mandate from the American Congress to negotiate in Seattle. Governments are represented at the UN, but not popular movements. The Seattle movement has its own legitimacy, but a permanent forum for such campaigns must be found to force governments to recognize them. I believe that the UN, with its numerous charters, is the only organization capable of integrating this idea of popular representation. A proposal for a second chamber at the UN has been put forward. I don't know if it's the best solution, but there must be a debate on the issue involving trade unions, farmers' movements, ecologists and consumer organizations.

KEEPING YOUR FEET ON THE GROUND

Could this be the beginning of a movement or a political party, the launching of a project that extends across the whole of society?

JB: All the issues we've discussed, put together, constitute a body of policy, but they're not the basis for a political movement or a party. No party has all the answers. This approach has been shown to be historically bankrupt, and we're not going to start it all over again.

The strength of this global movement is precisely that it differs from place to place, while building confidence between people. One of the lessons of Seattle was how this mutual confidence developed over eight days between groups holding very different – and, indeed, opposing – viewpoints. The sharing of experiences of struggle and solidarity will strengthen this burgeoning worldwide movement. A new citizen's power was born in Seattle.

Can this global movement begin to develop an anti-free-trade programme?

JB: The world is a complex place, and it would be a mistake to look for a single answer to complex and different phenomena. We have to provide answers at different levels – not just the international level, but local and national levels too. History shows that each phase of political development has a corresponding institutional form: France's response to the Industrial Revolution was the nation-state; the WTO is the expression of this phase of the liberalization of world trade.

José Bové, since summer 1999 you've been very much in the media all over the world. You've become an icon, a candidate for 'man of the year' polls: José this, José that. Doesn't all this coverage push away the very people you want to attract? Has José Bové been seduced by the world of showbusiness?

JB: We mustn't lose sight of the fact that the successful actions we took were within a trade-union framework. I belong to a collective movement; François and I weren't pulled out of a magician's hat. We have thirty years of struggle behind us, along with all our friends. When someone comes to the front of the stage – I did, after all, spend three weeks in jail – the media always personalizes the issues; that's just the way they are.

Attempts have been made to pull me in all directions, to get me involved in all sorts of different ventures. According to *Le Monde*, the Socialist Party even put out a rumour that I was going to stand for President! But we're not stupid. There is absolutely no risk of me being sidetracked into being a star. The media can say what they like – the Farmers' Confederation and our comrades in struggle will remain outside the system.

Are you going to concentrate your energies on work in Geneva?

FD: In the Farmers' Confederation, we fight on issues that affect other sectors. It's impossible to separate changes in agriculture from other parts of the economy – employment, for example. Agriculture is the first French professional sector where a union has not just fought for its own interests, but has questioned its own practices. It's a new phenomenon, and it's not surprising that it should originate with the farmers. Farmers' work, after all, affects everybody.

JB: I believe that the watchdog organization overseeing the WTO will not go unnoticed by the European Commission. We need to plan how we can bring Seattle to a local level in a way that will make our elected leaders face up to their responsibilities and stop capitulating to the interests of capital.

OPTIMISTIC FARMERS

You seem very sure of the outcome of your struggle . . .?

JB: Yes, and that's why we'll win. Our experience of struggle all over France – from our campaign for fairer land distribution through our opposition in the 1970s to the extension of the military camp at Larzac, to the recent fight against GMOs and junk food – has shown us that we can be successful.

On the issue of GMOs there have been huge advances, the latest in January 2000 when a protocol, agreed in Montreal by one hundred and thirty countries, adopted a precautionary principle on the import of GM produce. Other disputes have been less visible, but significant in developing awareness of what the stakes are. What's important is the educative value of an action – whether it encourages public participation. Actions that exclude people are failures. Actions can change the ideas of those who take part as well as those who observe them. Millau and Seattle showed the force of direct action. Legitimacy is a prerequisite. We had that on our side in Millau, with the ban on Roquefort. Similarly during the Larzac struggle. Often illegal action is required to make a case. If the case is fair, the public will support it. Action is collective, but responsibility has to be assumed individually – including going to prison, if that's necessary. In order to win, you have to be sure there will be solidarity with your action. In any case, if there's no hope of winning, there's no point in starting the fight.

And you, François . . .

FD: We've always associated our trade-union work with both thought and action. This is the *raison d'être* of the Farmers' Confederation: to situate problems in the context of society as a whole. We could very easily have stayed in a cocoon, quietly getting on with our alternative agriculture. But our predecessors' struggle, and our roots, have taught us to respond to society's wider concerns. Seattle was a historic event, marking the emergence of a new awareness on a world scale. For us, it means that our predecessors' struggle is bearing fruit today.

9

POSTSCRIPT:

SEATTLE-ON-TARN

Midnight in Millau, 30 June 2000: Loud clapping drowned the last chords of the band 'Noir Désir', whose music had filled the air. There were tears of gratitude in Bové's eyes as he approached the mike: 'If, today, we are able to say no to being seen as a commodity, it's all because of you.' An ovation rose up from the crowd. At least 100,000 people were present – maybe more. The Maladrerie stadium was packed; the banks of the river Tarn were full, as were the streets in the old city centre. People of all ages, from all walks of life. Lori Wallach, leader of Public Citizen, the American consumer organization set up by Ralph Nader, whispered to José Bové: 'It's wonderful. There are twice as many people here as there were in Seattle.' The Farmers' Confederation spokesperson pointed an accusing finger at the French presidency of the European Union: 'We're here to tell Chirac and Jospin: "You can't just do what you like without consulting us. We will not accept the selling of citizens' rights to the multinationals or the WTO. We are here to resist, to construct, to reclaim power at the base, and not let ourselves be manipulated by the world's powerful people".'

The crowd was addressed by the ten accused farmers, prosecuted by the state for their stand against McDonald's. The accused mocked their prosecutors, shouting: '*You* are the sentence,' and thanked the tens of thousands of people who stayed well into the night listening to their accounts of their trials, and to the artists and musicians who were performing at the rally.

The organizers of the demo had expected 30,000 people; instead, thousands of cars, four hundred coaches and special trains deposited huge numbers arriving from all over Europe: Italian, Belgian, English, German, Dutch and Swiss people had taken a few days' holiday to come and show their support for the accused.

The big surprise was the massive presence of young people. They were expected to turn up for the concert, but they had arrived at the start, on the Friday afternoon, to take part in the forums, walk around the stands, chat in the shade of the trees, stroll round the old town. Whether they were unemployed or steelworkers, philosophers or farm workers, they had come to demonstrate against a WTO capable of forcing nations to dismantle their legal systems, abolish social and environmental regulations, and insist that the Third World countries negotiate with it when the USA itself had still not signed the Charter for the International Organization of Labour. Some of those youngsters would go on to Prague in September, for the reunion planned during the IMF meeting there. Millau had become the latest stop on a road that stretched from Seattle to Bangkok, from Davos to Washington.

Although the local council had warned shop owners that they should pull down their shutters in case there was a riot, the only victims were those few who heeded this warning and lost sales because they were closed. Millau's population had grown fourfold, but not one incident occurred – and this despite the continued hostility of the mayor. At the end of the protest, hundreds of volunteers from among the demonstrators cleared up, so it was soon impossible to tell that two days of meetings and festivities had taken place.

The atmosphere was more like that of a big rally of many different interests rather than a demonstration. It was a fête, where a highly motivated new social movement was struggling to get to grips with the basics. The whole town was in festive mood. The support movement around the accused farmers was much wider than the traditional left. Militants from the Socialist Party, the Communist Party and the Greens attended alongside those who were present because 'the left has sold out'. Others there had never been involved before, but got involved on the issue of food, because 'you are what you eat'.

Those protesting against poverty stood side by side with trade unionists taking action against deteriorating working conditions. They have no faith in 'the Revolution': 'We've come to build a social movement that started in 1995. We're not a minority.' It was not so much a crowd, more a multitude of variously motivated individuals, pooling ideas in fourteen different forums.

The demonstrators in Millau knew there were no easy answers. They did not need to be told that all the different problems cannot be solved at once. They were looking not for a system to 'save' them, but for a range of solutions, to help reinstate their values. They were seeking diversity and a sharing of resources. They wanted a mature militancy, where people understood the need to build a worldwide network, with all the complexity that entails.

Of course, the farmers were there in force, and thrilled to be basking in the new-found support for a fight they had started alone. Their sacrifice of choosing to work outside the intensive farming system had at last been recognized. They, too, are pragmatic, and realize that 'in order to achieve an equitable globalization, the WTO's power must be addressed'. Their aim is to 'change politics gently' by putting their beliefs into practice: 'We try not to work too hard, so that others can have work.' They play the economy at its own game: 'By basing our purchasing on ethical grounds, we can say No to GMOs. We have the means to take citizen actions, to boycott.'

Do they believe they can change the course of the world? 'Look at the nuclear issue: in the 1970s, only a minority were anti-nuke. Since Chernobyl, this has grown. Twenty years ago, ecology meant nothing; today it's vitally important. History has always developed from minority movements which have then grown. In any case, the question should be asked after the event, not before.'

There was much discussion, but the trial was never far from people's minds. People were stationed by the windows of the courthouse, and the accused were accompanied by shouts of 'We shall win' when they went in – the only real slogan of those two days. Updates on developments were given from a podium set up in the square next to the courthouse, and defence witnesses from all over the world stopped by to repeat what they had said on the witness stand. Representing farmers' movements were Bill Christinson from the American National Family Farm Coalition (NFFC); Rafael Alegria, general secretary of *Via campesina*, from Honduras; Piotr Dabrowsky, ex-Minister of Agriculture in Poland; Rafael V. Mariano, representing the Philippine farmers; Mamadou Cissoko, leader of the CNCR, from Senegal; François Dufour, on behalf of the Farmers' Confederation.

Experts on globalization who testified included Susan George, political scientist, co-founder of ATTAC and Director of the new watchdog on globalization; Lori Wallach from 'Public Citizen' in the USA; Vandana Shiva from the RFSTE in India; Paul Tran Van Tinh, French ambassador and former European negotiator on GATT. Representatives of the trade-union movement included Louis Kotra Ureguei, from the Union of Workers' Unions and Exploited Kanaks; Hiro Tefarere, former general secretary of the Polynesian union A-Tia-I-Mia; Gilles Sainati, general secretary of the Magistrates' Union. Pierre Laur, a Roquefort industrialist, also attended; as did Paul Ariès, an academic and author of *Petit manuel anti-McDo*, who turned up dressed as the clown Ronald McDonald. Guy Durand, local opposition councillor, was there representing the original 12 August 1999 demonstration. Inside the court, under heavy police protection, the ten accused – José Bové, Jean-Paul Delaitte, Raymond Fabrègues, Gilbert Fenestraz, Frédéric Libot, Léon Maillé, Richard Maillé, Christian Roqueirol, Jean-Émile Sanchez, Alain Soulié – stood together with their defence lawyers: Marie-Christine and Christian Etelin from the Toulouse bar, Jean-Jacques de Félice and Henri Leclerc (former

President of the League for Human Rights) from the Paris bar, François Roux from Montpellier, and Maaroufa Diabira from Mauritania.

THE GLOBALIZATION TRIAL

You were both inside the court: José on trial and François as a witness. What was it like?

JB: It was simultaneously serene and intense, as if everything that had happened in the previous year was condensed. As we stood before the judge, the 12 August 1999 action and the debate on food, the issue of globalization and the WTO, were all dealt with in a few hours.

We knew that anything we said in court would be repeated to the crowd of supporters outside in the streets, so we had to be very careful with our words.

FD: I had difficulty imagining the atmosphere outside on the street. I spent six hours in the witness room cut off from the protesters and the courtroom. It was hot and sticky, and we were under great pressure all afternoon. I had been a witness once before, at the trial for the destruction of the genetic maize in Agen. I was on edge, because all the other witnesses were very nervous, especially those who needed an interpreter, and also because I was the only representative of the French farmers to give evidence. I was worried about what was happening outside, hoping that there would be no violence, especially at the evening rally. And I had to concentrate in order to give effective voice to the farmers' struggle.

Inside the courtroom, knowing how the judges had behaved in the previous weeks (we'd had to twist their arm just to accept our witness reports), I was worried that they would cut us short. But on the positive side, it was very reassuring to see all these witnesses who'd come from the four corners of the planet. With all that they had to say, I told myself, there was no way a discussion of globalization itself could be avoided.

When I entered the courtroom, I noticed that the accused were relaxed, confident and proud of the actions which had led them to the dock. I did worry a little about the judge, who was both animated and unpredictable. While I was under examination, I caught his eye, and what I saw was not reassuring. I had the impression that here before me was a man spoiling for a fight. I knew he was ill at ease, and had difficulty accepting the testimony of Gilles Sainati, who is also a judge, as well as being a witness for the defence.

In September 2000 I was again called upon as a witness – this time in Foix, at the trial of some mates who had pulled out genetic rape seed crops. As I speak, the verdict is not known, but the atmosphere at that trial was more relaxed, with the judge allowing witnesses to give confidential reports.

Was the presence of the demonstrators felt inside the court?

JB: Yes, the trial wouldn't have had the same meaning if we hadn't seen the big crowds assembling on our way in. There were two sides to the trial: while inside we were explaining ourselves to the robed representatives of justice, outside tens of thousands of people had encircled the courthouse and were pronouncing their judgement on the WTO. Inside, the judiciary was trying to reduce collective action to individual charges, while outside, citizens were exercising collective power to condemn an anti-democratic international institution.

FD: It was wonderful to have that relationship with the street. Every time there was an adjournment, I would go and have a quick look out of the window, and as far as the eye could see the streets were filled with crowds. There were so many journalists that it was suggested that a second room, with simultaneous transmission of the trial, should be set up. This was further proof that our actions and their legal consequences were being followed by the whole world. It gave me confidence, and for a moment I thought it would weigh decisively in our favour – that nine men would be acquitted and José would get the minimum: that is to say, a suspended sentence. I believed that the court would take into consideration the tens of thousands in the streets. Surely the judges were aware of the political significance of the movement since the McDonald's demonstration, and particularly after Seattle? I was convinced that the legitimacy of the Millau action would be recognized.

After all, it had decisively helped in reducing the risk of our eating hormone-fed meat and GM foods.

Is it customary for judicial proceedings to be influenced by mobilization on the streets?

JB: I'm not under the impression that the street had any influence on the course of justice. The fact that both the trial and the assembly outside occurred alongside each other was in our favour, but at no point did the judge have to ask the demonstrators to be less noisy, or call on the police to intervene. Those who had come to support us were very much aware of what was going on, and had no wish to disrupt the proceedings.

FD: It was a trial unlike any other. The destruction of GM crops, and the dismantling of McDonald's, were not corporatist actions carried out purely in the interest of farmers, but actions affecting the whole of society. As consumers and citizens themselves, the judges would surely bear this in mind, and pass sentence accordingly.

Judges tend to be very touchy about their independence, and certainly don't like having their arms twisted. Don't you think that maybe the massive crowd had a negative effect on the severity of the sentence?

JB: I don't believe the presence of the crowds had any influence on the sentences. The judges would have had to consider the broader context of the issues and the public outcry on food, Roquefort, Customs surcharges and GMOs anyway. The large turnout in Millau was evidence of the resonance of our action. But it was not the intention of those assembled to twist the judges' arms. They were there to show that another trial was taking place out on the streets: the trial of the WTO and junk food.

The defence was not confident that all the witnesses would be heard. How did they succeed in convincing the court to let them all take the stand?

JB: As soon as the date of the trial was set, Judge Mallet let it be known that we would be tried during the normal Millau court session, in between other ongoing cases. It was his way of belittling our action, dismissing its uniqueness. As time went by, the magistrates had to change tack. First they talked about setting aside an afternoon for our case. Then it was agreed that the trial would have to be held over two days if the issues were to be examined in depth, and all our sixteen witnesses and six lawyers were to be heard. The judge now understood that it was not possible to reduce our trial to one of common law, hence the increase in the allotted time; but the verdict did not reflect the content of the debate. . . .

FD: In the weeks running up to the trial, lawyers and supporters had argued that the judges should take into account the wider movement against globalization. But given the rigid approach of the judges in the August 1999 trial, putting globalization in the dock did not guarantee us a favourable verdict.

How did you prepare your defence?

JB: The same way as we did everything else: collectively, clearly reiterating that everyone was equally responsible. There were no leaders or followers, only trade unionists and committed individuals.

From the start of this whole affair, we had called for an acquittal. Our lawyers pleaded for this as the only answer to people who had taken symbolic action to defend a legitimate cause, even if this action was technically illegal. Henri Leclerc summed up for the defence by reminding the court of previous farmers' struggles, including those of wine growers in the South of France, which had expressed an understandable anger with serious consequences, yet had not brought official retribution.

FD: For the Farmers' Confederation, mobilization had occurred in stages, from McDonald's to the trial. There was Seattle, then meetings up and down the country building mass support for the farmers. Hence the importance of using witnesses from other sectors and countries. With these witnesses, and public opinion, we had to prove to the court that the actions taken were those of a movement that was neither corporatist nor typically French. From the start, the union had been mobilized on the issue of globalization, not just our particular struggle, linking ideas and actions with other social sectors.

Were there any amusing incidents during the trial?

JB: When the judge presented photos of the dismantled McDonald's supposedly proving the damage done, all they showed was a construction site with some graffiti here and there, not the ruins of a ransacked building. The impact on those in the public gallery and journalists who were unfamiliar with the place was immediate. They accepted our version: that the action had involved a dismantling rather than a wrecking.

Another surprising incident was when the judge discussed the order of witnesses with his colleagues. The microphone was still on, and he was heard to say that Susan George should be last, because he wasn't going to let himself 'be pissed off by some leftie'. That was very surprising to anyone who knows Susan George.

There were also moments of tension when the police officers and detectives came to the witness stand to say they'd heard me threaten to plant a bomb in any future anti-McDonald's action. Under cross-examination from our lawyers, they began to look uncomfortable. The officers stuck to their story, but the town police declared they could not remember me making this statement – and rightly so. The court did not charge me for this, but the sentence was still three months.

FD: Yes, we all had a bit of a laugh when the photos of the building site were passed around and the McDonald's representatives, even though they had not brought any charges, had the nerve to bring along a bill for 700,000 francs, without any assessment of the damage. As he was questioning me, the judge was leafing through the pages of the original version of *The World is Not for Sale*, and admitted in full session that he'd learned a lot from reading it, especially about GMOs.

The screwdriver incident also caused some amusement: speaking as if a bulldozer had driven over the site, the judge said to the accused: 'You were in possession of a screwdriver and a crowbar, and you unscrewed a panel; that's a real act of violence. You contributed to the 700,000 francs' worth of damage.' He also spent some twenty minutes wondering where one of our friends who'd been accused of spray-painting a panel had got the paint from. . . . A long way from the subject of hormone-fed beef! The lawyers had to rescue the judge and urge him to move on from such ridiculous matters.

Looking back, what was the high point of those two days – the one thing that immediately springs to mind when you think about it?

FD: My first mental image is my recollection of that evening: the overwhelming feelings of happiness and nervousness as I got up on the stage with Philippe Val. The floodlights illuminated the crowds, and as far as I could see there were people, and yet more people. I had tears in my eyes when I spoke into the microphone: 'We were expecting 4,000 and there are 100,000 of you.' Every time I think of Millau, I see those shining crowds.

I have another vivid memory: our journey from the farm to the court on the tractor trailer was like a trip to the scaffold. All along the way, people kept coming up to us, shaking our hands, giving us encouragement.

JB: I have the same memories of that evening. We, the ten accused, and François mounted the podium, and after greetings I began to speak to the 100,000 people who had come to support us. It was a powerful moment. I just wanted it to continue, to suspend time, so that I could savour the moment to the full. François and I were both moved to tears.

AN UNACCEPTABLE VERDICT

The verdict was announced on 13 September 2000, and the sentences were three months' prison for Bové; two months' suspended sentence for Jean-Émile Sanchez, Richard Maillé, and Frédéric Libot; a 3,000 franc fine for Jean-Paul Delaitte, 2,000 francs for Christian Roqueirol, Raymond Fabrègues, Léon Maillé and Alain Soulié; and Gilbert Fenestraz got nothing. Were you surprised by the judgement?

JB: No, I expected that the judge would want to set an example. All the trade-union groups, with the exception of the FNSEA, have condemned the sentence. The court set itself up as a defender of law and order, as was clear from the reasoning given for my three-month prison sentence: 'A sentence to make him see reason. . . .' With a sleight of hand, the judge dismissed our argument that the only way to make ourselves heard was to take direct action.

FD: If the judge's intention was to break the solidarity of the team and isolate the militants by handing down harsh sentences, he failed. But he didn't take his decision alone. It reflected the attitude of the local government of Millau when it tried to intimidate the local population on the eve of the trial. The judge had failed to acknowledge the strong body of opinion on the issue. But I'm not surprised by the outcome. I'm thinking of other trials of Farmers' Confederation militants in the area, particularly one relating to union representation in April 1995. When some Farmers' Confederation comrades wanted to attend a meeting to discuss the agri-environmental aid budget, the FDSEA pressurized the Department's agricultural and forestry representatives to bar them, even though they had the legal right to be represented. The meeting was cordoned off by the CRS, who were only too happy to have a go at our members; while inside, the FDSEA was deciding the budget allocation. This incident is illustrative of how the law can be stretched by those who are supposed to implement it, and how the administration is in the pocket of local power.

Anyway, the verdict has given us the chance to appeal. We will bring even more people together and establish links with other trials, as well as the role of the WTO. Australia and New Zealand have refused to set a new list of dates for WTO meetings. They believe such events are doomed, as there has been no change in attitude since Seattle. We have obviously had an impact, and we intend to broaden the appeal trial to include all struggles against blind globalization.

But isn't the law the same for everyone? Why should farmers who have committed a crime be treated more leniently than young people from the suburbs who get heavy sentences for minor offences?

JB: You can't judge this in the abstract. The context and motives of any action have to be considered before deciding whether or not it's criminal.

FD: Analysing motives is the judge's responsibility. He has to decide between intentions, actions, and the law. The court had sufficient explanations on the motive of last resort for our action. We had put our case to the Ministry for Agriculture, to the Prime Minister's office and to the European Commission. When no one listened, what could we do but take action? And it's by no means the end of the story. The Americans have just confirmed the continuation of the surtax on some European products, and suggested that the list will be lengthened.

I'd like to ask publicly whether the courts are competent to judge problems of this kind, of such international significance. A country judge, sitting alone, is not competent to provide the answers to the havoc created by the global market. There appears to be a widening gap between justice and society.

Cases like this make me wonder whether we need more experts, such as sociologists, to help the judges and the public understand what's going on. This could also apply to the cases of young suburban delinquents who go off the rails.

Can you talk about criminalization of a social movement?

JB: The most significant fact is what happened with RAGT (Monsanto had been carrying out trials for the company RAGT in Druelle, in the Aveyron) when, after the destruction of a small plot of genetic maize, a judge put the Farmers' Confederation itself on trial, even though the eighty participants in the action had handed the police a list of their names. For the first time in France's history a trade union was on trial, with the ultimate threat of being dissolved. We were asked, again, to name a representative of the union – its moral authority, so to speak. In the absence of any response, the judge nominated a representative.

FD: Agen, Foix, Millau, Druelle, but also the sentencing of militant workers such as Michel Beurier. . . .[1] We are witnessing a worrying criminalization of

1. Michel Beurier: trade-union leader and activist, CGT representative for Puy du Dôme (Massif Central). Arrested and tried for action against GM crops.

social movements. The court in Millau showed that it was willing to hold José Bové responsible for everything – as if our movement was divided into gurus who think and minions who work. This approach smacks of anti-unionism.

Does the idea of independent justice have any validity?

FD: We hear a lot about independent justice, and the principle is good, but when our elected representatives remain silent on important questions, it's up to ordinary people who understand what's going on to move into action. At this point they often run into conservative judges. It's a dangerous situation, as the politicians can't stand aside and watch the judiciary take a beating. The citizens and the judiciary can't be left face to face, alone, without the intervention of the politicians.

Furthermore, some questions need to be asked. The Ministry for the Environment was ransacked, the Town Hall in Brest was destroyed, public buildings were attacked – evidently by some bunch of corporatist agricultural workers, but no one bothered to pursue them. On the other hand, heavy sanctions are heaped on those who take legitimate action, symbolic action that is in line with public feeling. Another example: an RPR treasurer is given a nine-month suspended sentence for diverting considerable sums of public money, and José is given three months . . . where's the logic in that?

In Millau, an issue of international concern was judged by French law and local petty-mindedness. Does justice in our country still reflect the wishes of the people?

A NEW CONSCIOUSNESS

How do you explain the importance of the mobilization?

FD: The actions of the Farmers' Confederation were recognized as a legitimate defence. Between the McDonald's dismantling and the trial, we witnessed the rise of a new awareness on a world scale, confronting issues hitherto dealt with only by experts. The coincidence of dates was in our favour: if Seattle hadn't happened between the Millau action and the trial, maybe we wouldn't have had the same turnout. It was in fact José, on his release from prison, who set the ball rolling: 'We have to go to Seattle. That's where the next event is.' It was José who said that, not the Farmers' Confederation. After Seattle, we followed up with the anti-Davos protest and a series of meetings. . . . Whereas in the early 1990s people were unaware of the WTO, now, in the space of ten months, the whole world became interested in the deals that were being cooked up in the offices of its leaders. People woke up. The strong anti-WTO movement in India, the mobilizations in a number of countries against the AMI, all occurred because people understood what was lurking behind the opacity of these international institutions.

From the moment that governments or political parties calling themselves democratic were no longer able to reassure people or give them answers – that was when people started to take direct action. Their common aim, despite any political differences, is to combat the standardization of the world. This explains the importance of the Millau rally, and the presence of so many 'unorganized' people.

Another remarkable issue raised by Millau is the coming together of veterans of the struggle for the Larzac with young people, twenty-five-year-olds, who found out about that campaign only in the discussions after the Millau events. Veterans of the anti-militarist fight are used to telling the youngsters about the glories of their old battles. Now the twenty-somethings have joined the struggle themselves. Their parents mobilized over food, but these young people woke up on more overtly political issues. Dismissing the traditional parties as useless, they felt that it was important to go to Millau, because they saw a ray of hope. I have been astounded by the number of occasions in the wake of Millau when young militants have demanded better working conditions or salaries from their bosses.

JB: As François has explained, people mobilized for more than defence of trade-union rights, or for wholesome food, or against the Americanization of justice. They mobilized against a society at risk. We see the emergence of another dimension to the mobilization which can be explained by the threat globalization poses to a fundamental tenet of society, food self-sufficiency. Hence the demand to be allowed democratically to choose one's agricultural crop, and to be supported in this by state funding. With the gradual reduction in working hours in post-industrial society, the search for food is no longer a priority for most people, and its cost isn't their biggest expense.

A variety of 'unorganized' people came to support you for a host of different reasons. According to many trade-union workers, they ran into comrades they hadn't seen for a long time. Does Millau represent the beginning of a new militancy?

FD: I met a lot of people in Millau who admitted they had not been active for twenty years. They were not proud of this, but the struggle is giving them renewed confidence in the possibility of changing things. They were waiting for a sign. We're getting lots of phone calls and letters thanking us for organizing the Millau event. The support movement has given hope to a great number of people; hence the plethora of local and international rallies. People have started to look out for any international meeting where there's a chance of decisions being made against the public interest.

But how do you direct this sentiment?

FD: People proceed on the basis of what's happening worldwide; each new rally holds out hope, is proof that the worldwide challenge is being maintained. Seattle, Washington, Davos, Bangkok, Millau, Prague, Bangalore . . . these didn't happen in one day. Since Millau, calls are received every day at the ATTAC offices requesting local contact numbers. People want to get in touch with others who have the same outlook. It's an odd sort of militancy, with no specific political project. But people know what they don't want, and that's a good sign. There's no relation to what used to happen in the past, when people would assemble in their professional fiefdoms – students in their universities, farmers discussing the price of their products, and so on. Today, new links have been created.

JB: It's a question of being militant in a different way, by understanding that interdependency is axiomatic. By rallying as many consumers as farmers, the demonstration of 12 August 1999 brought collective interests back to the forefront. After a year of debates ranging from what's on our plate to the future of globalization, in a society where the production process has atomized the producer – be it in a factory or in the countryside – people have felt the need to come together, to be counted, to join others who believe it is still possible to change the world. The number of requests from non-farmers to join the Farmers' Confederation is further evidence of what's happening.

A hundred thousand people mobilized, more through the networks . . . a huge movement, but with no desire to take power. Where does this lead?

JB: A hundred thousand people means a hundred thousand individuals, each responsible for their own actions, each able to do things in their own struggle. Today, people mobilize without wanting to take over state institutions, and maybe this is a new way of conducting politics. The future lies in changing daily life by acting on an international level. This way of mobilizing corresponds to a new awareness of the economy's growing independence from political power. The multinationals take decisions with complete disregard for nation-states, displaying contempt for the political system. That requires new responses, new forms of militancy. This is what happened in Seattle, Millau, Prague, and elsewhere. The preconceived model of how to change the world has been replaced by a new form of collective action, and new organizational structures are emerging.

But can militant trade unionism regenerate or reinvent politics by demanding of the traditional political leadership that they assume their responsibilities?

FD: People no longer expect a 'revolution', but they do want to influence the course of events. They've drawn the conclusion, after much debate, that the problem exists at both a local and a global level. Genetic pollution doesn't stop at national borders.

JB: While we're still waiting for the organizational form that this opposition will take, its legitimacy is endorsed with every world rally, and goes beyond the traditional parties. Politics is given a new credibility. Globalization is occurring as rapidly from the base as from the top, thanks to the movement.

In France, where farmers represent barely 2 per cent of the population, it's not surprising that they have sought unity with others. They were almost wiped off the map, politically and socially, and suddenly, in the last year, they are at the forefront of political debate. How come?

JB: Because farmers, unlike the 'pig MPs' and 'wheat senators', have not betrayed the confidence of the consumer. These farmers warn against the practices of intensive agriculture. Consumers have heard the call, and responded.

FD: There was no alternative but to react when we were confronted with the disappearance of our jobs. Forbidding us to use our own seeds, or forcing us to use hormones, pushed us into the leadership of a huge movement.

The alarm was raised by a society of sixty million people, only a million of whom are active in agriculture. Many countries in the world still have far larger percentages of farmers. Does the concept of 'wholesome food' have a meaning in the South?

FD: The dominant model wants to introduce GMOs. We know this is not a satisfactory response to famine: it standardizes produce; its objective is to obtain profit; nutrition is not the issue. Generally we eat three times a day: this is a right for every citizen . . . and a captive market which the food multinationals have grasped. The best way to monopolize this market is by removing the national autonomy of individual countries.

It is possible to mobilize more people in developing countries, because the farmers there are more numerous. An example: last week, a doctor in my area came to ask me for a contact with *Via campesina*; he has a friend in Brazil, part of the Ameriindian community. We're talking about some two hundred families who are reappropriating land and looking for help in cultivating local plant varieties that are under threat. I've had six requests of this sort since Millau.

JB: Our witnesses at the trial emphasized the importance of wholesome food in the South. Food self-sufficiency, its quality and safety, are as important in the Third World as they are for us. This is the vital link bringing all farmers and citizens together: a real globalization, which gives us hope.

I went to Bangalore to meet the Indian farmers waging a struggle against the multinationals who are trying to impose GMOs on them. Farmers in developing countries reject GMOs as an alleged solution to hunger. What better evidence could there be that our fight is justified, and that it also encompasses countries of the South?

In the world today, twenty-seven million farmers work with a tractor, 250 million use animals, and a billion work with their hands – the latter almost exclusively women. So when we're considering the development of agriculture, the first issue is to determine how all these people can continue their work in a way that is satisfactory to them.

GLOBALIZING HUMAN RIGHTS

Seattle had said 'No' to the WTO, and, as Henri Leclerc stressed, Millau said 'Yes' to the internationalization of human rights . . .

JB: In calling for the acquittal of the ten accused in Millau, innocent public opinion also demands that social and human rights are applied internationally. Henri Leclerc pleaded very effectively for this in court. In Seattle, the Farmers' Confederation marched behind a banner calling for trade laws to be subordinated to human rights and the main UN Charters.

FD: Henri Leclerc was right; the market has abolished frontiers, and seeks to impose uniformity on the planet. It's up to us, as citizens of the world, to raise the question of rights for everyone. Human rights don't stop at frontiers; we must globalize them.

José, where did you get the idea of asking the 100,000 people who were present on the evening of Friday 30 June to clap and chant 'Liberty, Equality, Fraternity'?

JB: The idea came talking to people, from seeing what was happening and what was being said, from the warm atmosphere throughout the rally. If someone trod on your toes, there were apologies; and in the forums, the debates were friendly, never abusive. The word 'fraternity' seemed like a natural description.

Freedom, too, is what we're calling for, especially trade-union freedom, which we refused to buy through paying bail. And equality means decent living conditions and a chance to develop for all.

These were the ideas that were going around. That evening, at the foot of the stage, I was chatting about the day's events with the designers of the magazine *Charlie Hebdo* and others, when one of them drew a picture of a *sans-culotte*, the name for the poor in the French Revolution. That's what made me think of the slogan 'Liberty, Equality, Fraternity'. The struggle of the *sans-culottes* was no different from ours today.

The victory of the *sans-culottes* was appropriated by the stooges of the state, so that the slogan 'Liberty, Equality, Fraternity', painted on the façade of every town hall, has lost its meaning. It's up to us, in the streets, to rehabilitate democracy from the grass roots, to take over the slogan and give it back its original meaning.

As I was about to say the words, another image came to me. It was of the American trade unionists, arrested after the bloody Chicago days of 1st and 2nd May 1886, when the police opened fire on demonstrators demanding an eight-hour day. Their leaders, who were sentenced to death and executed, were of different nationalities, and when they got up on the scaffold, they all started singing the *Marseillaise*. It was the only song they all knew, and they wanted to die defiantly. Their death is marked by May Day celebrations around the world. Incidentally, the workers in question worked for McCormick, one of the first manufacturers of agricultural equipment.

At the close of the rally, José called for people to 'Think globally, act locally', and demanded that the fight against GMOs should continue. Today in France, GMOs are being mixed surreptitiously with normal seeds. What does acting locally mean in this situation?

JB: Thanks to our campaign, the French government was forced to destroy 600 hectares of genetic rape seed. But they're still dithering about 5,000 hectares of maize polluted by GMOs. Apparently they're unconcerned about the accidental 'mixing' of traditional seeds and GM seeds under the pretext of acceptable thresholds. But there's no acceptable threshold for genetic pollution. If the state won't destroy GM fields, then we will.

FD: Today, we're in the firing line. We must refuse to give in, and continue to prevent all genetic planting through our mobilizations. That is the best support for those standing trial accused of destroying GM crops. We also need to find new forms of activity that can have an impact on the scientific community, which is currently very ill at ease. I noticed this at the trial on the destruction of genetically modified rape seed in Foix. Although several scientific witnesses accused us of destroying their work, outside the court-room they admitted to me that they understood our position. A number of researchers have asked for discussions with us. I believe we should make the scientists face up to their responsibilities. They are, after all, citizens; they have children; they need to consider future generations. Their scientific behaviour lacks an ethical dimension.

Would you consider organizing a big symposium on GMOs, bringing together farmers, scientists and consumers?

FD: I'd certainly be in favour of that, as long as the principle of international participation was accepted right from the start, including developing countries.

You've suggested the idea of an annual meeting on 30 June, making this the anniversary of the struggle against the multinationals. Do you still stand by that?

JB: Rather than meeting at a set location on the anniversary, 30 June and 1 July could become dates for different get-togethers in defence of the people against the interests of the multinationals. On 13 September 2000 we were sentenced for our action in Millau. We have decided that the next rally will be on the date of the appeal, in Montpellier, and we're calling for a big turnout.

FD: I'd like 30 June or 1 July to become a date for meetings in those areas where struggles are taking place. The start of the summer break could be the time to get together and combine festivities and debates on the experiences of the year gone by. With the sentences in Millau, meetings in 2001 could discuss the criminalization of social movements, including a debate with representatives of the judiciary. We want to force a change in the way the judicial system operates, and we've shown that we're capable of putting a great many people on the streets to make that happen.

AFTERWORD TO THE PAPERBACK EDITION: JOSÉ BOVÉ INTERVIEWED IN *NEW LEFT REVIEW* *

A FARMERS' INTERNATIONAL?

What was your formation as an activist in France — were you too young to participate in 1968?

I was then in my first years of secondary school, outside Paris, but of course I was affected by what was going on — the May events, the discussions, the whole atmosphere. I didn't do much, apart from an occupation of the school football pitch. It was in the last years at school that I started going on demonstrations. When I was seventeen I got involved in the struggle against military service — for the rights of conscientious objectors and deserters. There was a network of groups throughout France. We used to attend the

* This interview was first published in *New Left Review* 12, November–December 2001, and is reproduced here by kind permission of the journal's editors.

military tribunals every week to offer support for the boys doing military service – and for the regular soldiers, put on trial for stealing or getting into conflict with an officer. We collected all the statistics and publicized what was really going on inside the army. In 1970, '71, I moved to Bordeaux with my parents, just after the Baccalauréat. I had been born there, but my parents – agricultural researchers, who worked on the diseases of fruit trees – moved round quite a lot. We spent a few years in Berkeley when I was a child.

I could have gone to university in Bordeaux, but I wanted to work full-time with the conscientious objectors. It was then, in the early seventies, that the peasants of the Larzac plateau got in touch with us. The Army had decided to expand the military base there – from 3,000 hectares to 17,000. The local farmers asked for our support in setting up resistance groups. We built up a network of over two hundred Larzac committees in France; there were some in Germany and Britain, too. All new construction on the plateau had been forbidden so, in 1973, we started building a sheep barn there, right in the middle of the zone that the Army had earmarked. Hundreds, even thousands came to help – we called it a *manifestation en dur*: a concrete demonstration. We built it completely in stone, in the traditional way. It took nearly two years. At the same time, our network was in touch with a mountain farmers' group in the Pyrenees. We used to take military-service objectors to work up there, on land that's too steep and mountainous for machinery – everything has to be done by hand. That was where I had my first experience of dairy farming and cheese-making. Then, in the winter of '75–'76, the Larzac farmers decided we should squat the empty farms that the Army had bought up around the base. I moved into Montredon, as a sheep farmer – with many close contacts in the region.

What were the main influences on you at that stage?

There were two strands. One was the libertarian thinking of the time – anarcho-syndicalist ideas, in particular: Bakunin, Kropotkin, Proudhon, the anarchists of the Spanish Civil War. There were still a lot of Civil War veterans living in Bordeaux, and we used to have discussions with them. The other was the example of people involved in non-violent action strategies: Luther King and the civil rights movement in the States; César Chávez, the Mexican farm-worker who organized the Latino grape-pickers in California. There was a strong Gandhian influence, too: the idea that you can't change the world without making changes in your own life; the attempt to integrate powerful symbolic actions into forms of mass struggle.

In much of Europe and the United States, there was a clear rupture between the struggles of the sixties and seventies and those of today, with big defeats – Reagan, Thatcher – lying in between. In the States, in particular, there seems to be a new generation involved now in the anti-globalization protests. In France, there has perhaps been less sense of a clear-cut defeat, but less generational renewal, too?

The seventies were years of powerful militancy in France, coinciding with a political situation in which there was a possibility of the Left parties taking office for the first time. There was a lot of hope in 1981, when Mitterrand was elected. The ebb came in the eighties. Some people argued, 'We mustn't do anything that would damage the Socialists'. Others were disillusioned and quit politics, saying: 'We thought this would change things, but nothing has changed'. They were the years of commercialization, of individual solutions, when cash was all-important. We weren't affected by that so much in the peasants' movement. On the Larzac plateau, after our victory against the army in '81, we started organizing for self-management of the land, bringing in young people to farm, taking up the question of

Roquefort and intensive farming, fighting for the rights of small producers, building up the trade-union networks that eventually came together in the Confédération Paysanne. So for us, the eighties were very rich years. There was no feeling of a downturn.

As for the young generation: it's true that many of the campaigns of the nineties were a bit drab. They made their point, but they did not draw many people in. It was the emergence of another set of issues – the housing struggles of the homeless, the campaigns of the *sans-papiers* – that began to create new forms of political activity, crystallizing in the anti-globalization movement of the last few years. At the trial over dismantling the McDonald's in Millau in June 2000, we had over 100,000 supporters, lots of them young people. Since then, in Nice, Prague, Genoa, there has been a real sense of a different sort of consciousness. It comes from a more global way of thinking about the world, where the old forms of struggle – in the workplace or against the state – no longer carry the same weight. With the movement against a monolithic world-economic system, people can once again see the enemy more clearly. That had been a problem in the West. It's been difficult for people to grasp concretely what the new forms of alienation involve, in an economy that has become completely autonomous from the political sphere. But at the same time – and this may be more specific to France – the anti-globalization movement here has never cut itself off from other social forces. We've always seen the struggle for the rights of immigrants and the excluded, the *sans-papiers*, the unemployed, the homeless, as part of the struggle against neoliberalism. We couldn't conceive of an anti-globalization movement that didn't fight for these rights at home.

You founded the Confédération Paysanne in 1987. What is its project?

Firstly, it's a defence of the interests of peasants as workers. We're exploited, too – by the banks, by the companies who buy our produce, by the firms who sell us equipment, fertilizers, seeds and animal feed. Secondly, it's a struggle against the whole intensive-farming system. The goals of the multinationals who run it are minimum employment and maximum, export-oriented production – with no regard for the environment or food quality. Take the calf-rearing system. First the young calf is separated from its mother. Then it's fed on milk that's been machine-extracted, transported to a factory, pasteurized, de-creamed, dried, reconstituted, packaged and then, finally, re-transported to the farms – with huge subsidies from the EU to ensure that the processed milk actually works out cheaper than the stuff the calves could have suckled for themselves. It's this sort of economic and ecological madness, together with the health risks that intensive farming involves, that have given the impetus to an alternative approach.

The Right has always tried to control and exploit the farmers' movement in Europe, in accordance with its own conservative, religious aims. The agricultural policy of the traditional Left was catastrophic, completely opposed to the world of the peasants in whose name it spoke. We wanted to outline a farming strategy – autonomous of the political parties – that expressed the farmers' own demands rather than instrumentalizing them for other ends. We're committed to developing forms of sustainable agriculture, which respect the need for environmental protection, for healthy food, for labour rights. Any farmer can join the Confédération Paysanne. It's not limited to those using organic methods or working a certain acreage. You just have to adhere to the basic project. There are around 40,000 members now. In the Chambres d'Agriculture elections this year we won 28 per cent of the vote overall – and much more in some *départements*. It was 44 per cent in Aveyron, and 46 per cent in La Manche.

How did this come to pit you against the junk-food industry – most famously, dismantling the McDonald's in Millau?

During the eighties we built up a big campaign in France against the pressures on veal farmers to feed growth hormones to their calves. There was a strong boycott movement, and a lot of publicity about the health risks. Successive Ministers of Agriculture were forced to impose restrictions, despite heavy lobbying from the pharmaceutical industry. At the end of the eighties the EU banned their use in livestock-rearing, but it has been wriggling about on the question ever since. In 1996, the US submitted a complaint to the WTO about Europe's refusal to import American hormone-treated beef – exploiting the results of a scientific conference, organized by EU Commissioner Franz Fischler, that had concluded, scandalously, that five of the hormones were perfectly safe. But there was so much popular opposition, linked to people's growing anxieties about what was happening in the food chain – mad cow disease, Belgian chickens poisoned with benzodioxin, salmonella scares, GMOs – that the European Parliament actually held firm. When the WTO deadline expired in the summer of 1999, the US slapped a retaliatory 100 per cent surcharge on a long list of European products – Roquefort cheese among them. This was a huge question locally – not just for the sheep's milk producers, but for the whole Larzac region.

When we said we would protest by dismantling the half-built McDonald's in our town, everyone understood why – the symbolism was so strong. It was for proper food against *malbouffe*, agricultural workers against multinationals. The actual structure was incredibly flimsy. We piled the door-frames and partitions on to our tractor trailers and drove them through the town. The extreme Right and other nationalists tried to make out it was anti-Americanism, but the vast majority understood it was no such thing. It was a protest against a form of food production that wants to dominate the world. I saw the international support for us building up, after my arrest, watching TV in prison. Lots of American farmers and environmentalists sent in cheques.

How have you coordinated international solidarity with peasants and farmers in other lands?

From the early eighties, we started thinking about organizing on a European level. We felt we shouldn't stay on our own in France when there were other farmers' networks in Switzerland, Austria, Germany. We needed a common structure in the face of European agricultural policy, which is completely dominated by the interests of agribusiness. That was why we decided to set up the Coordination Paysanne Européenne, with its office in Brussels. It was through this movement that we got in contact with peasants' groups in other continents. It was about ten years ago that the idea of setting up an international structure was born. This was Via Campesina. There are many different peasants' organizations involved: the Karnataka State Farmers' Association from South India, which has played a big role in militant direct-action campaigns against GM seeds – they represent some 10 million farmers; the Movimento Sem Terra in Brazil, who lead land occupations by peasant families, and have an important social and educational programme. There are regional networks in every continent, organizing around their own objectives – Europe, North America, Central and South America, Asia and Africa. And then there is an overall coordinating executive which is based in Honduras at the moment, but will be moving to Asia next year.

You went to Seattle with Via Campesina. What was your critique of the WTO?

It was a big victory for agribusiness when food and agriculture were brought into the GATT process in 1986: a huge step towards regulating agricultural trade and production along neoliberal lines. Countries were no longer free to adopt their own food policies. They were obliged to lower tariffs and take a percentage of imports – which means, effectively, US and EU products: 80 per cent of world food exports come from these two. The process was taken further with the 1994 Marrakesh agreement that set up the WTO. Now a state can only refuse to import agricultural or food produce on the grounds of protecting the health of its population and livestock. The threat to these is determined by the Codex Alimentarius, which is in turn run by the food giants: 60 per cent of its delegates are from the EU and US.

The Marrakesh accords were supposed to be subject to a balance sheet at Seattle – of course, this never came. Not that we need an official report to know that the countries of the South have been the biggest losers: opening their borders has invited a direct attack on the subsistence agriculture there. For example, South Korea and the Philippines used to be self-sufficient in rice production. Now they're compelled to import lower-grade rice at a cheaper price than the local crops, decimating their own paddy production. India and Pakistan are being forced to import textile fibres, which is having a devastating effect on small cotton farmers. In Brazil – a major agricultural exporter – a growing percentage of the population is suffering from actual malnutrition. The multinationals are taking over, denying large numbers of farming families access to the land and the possibility of feeding themselves.

What were your demands at Seattle?

Firstly, all countries should have the right to impose their own tariffs, to protect their own farming and food resources and maintain a balance between town and countryside. People have a fundamental right to produce the food they need in the area where they live. That means opposing the current relocation of American and European agribusiness – chicken and pig farms, and greenhouse vegetables – to countries with cheap labour and no environmental regulation. These firms don't feed the local people: on the contrary, they destroy the local agriculture, forcing small peasant-farming families off the land, as in Brazil.

Secondly, we have to take measures to end the multinationals' dumping practice. It's a well-established tactic used to sweep a local agriculture out of the way. They flood a country with very cheap, poor-quality produce, subsidized by massive handouts in export aid and other help from big financial interests. Then they raise prices again, once the small farmers have been destroyed. In sub-Saharan Africa, livestock herds have been halved as a result of the big European meat companies flooding in heavily subsidized frozen carcasses. The abolition of all export aid would be a first step towards fair trading. The world market would then reflect the real cost of production for the exporting countries.

Thirdly, we absolutely refuse the right of the multinationals to impose patents on living things. It's bio-piracy, the grossest form of expropriation on the planet. Patents are supposed to protect a new invention or a new technique, not a natural resource. Here, it's not even the technique but the products, the genetically modified seeds themselves, that are 'patented' by half-a-dozen chemical companies, violating farmers' universally recognized right to gather seed for the next year's harvest. The multinationals' GM programme has also been a ferocious attack on biodiversity. For instance, something like 140,000 types of rice have been cultivated in Asia, over the centuries. They've been adapted to particular local tastes and growing conditions – long-grain, short-grain, variations in height, taste, texture, tolerance of humidity and temperatures, and so on. The food companies are working on five or six strains, genetically modified for intensive, low-labour cultivation, and imposing them in areas of traditional subsistence farming. In some Asian countries – the Philippines and China are the worst cases – these half-dozen varieties now cover two-thirds of rice-growing land.

What would be your alternative to the WTO?

We've argued for an International Trade Tribunal – in parallel to the International Court of Human Rights – with a Charter, and judges nominated by the UN. There should be transparency of action, and private individuals, groups and trade unions should be able to bring cases, as well as states. The Tribunal would play a constitutional role, advising on whether international economic accords should be ratified: they would have to concur with the individual and collective rights to which UN members are signatories – the right to food, to shelter, to work, education, health. These rights need to be imposed upon the market; they should be respected not just by states but by economic institutions. It's a similar process to that of the Kyoto accords on the environment.

Kyoto surely doesn't offer a very powerful precedent?

I agree. But these things take time. The call for an International War Crimes Tribunal has now been ratified by thirty or forty countries, although it's taken almost four decades. But it's essential to ask what structures we do want, for multilateral trade. We have to develop a long-term global vision, without being naïve. That will require a certain balance of forces.

Others in Via Campesina – the MST, for instance – have called for the abolition of the WTO, rather than its reform. Are the experiences of North and South at odds here?

'Food out of the WTO' is Via Campesina's demand. We're all agreed on the three main points – food sovereignty, food safety, patenting. For the people of the South, food sovereignty means the right to protect themselves against imports. For us, it means fighting against export aid and against intensive farming. There's no contradiction there at all. We can stage an action in one part of the world without in any way jeopardizing the interests of the peasants elsewhere, whether it's uprooting genetically modified soya plants with the Landless Movement in Brazil, as we did last January, or demonstrating with the Indian farmers in Bangalore, or pulling up GM rice with them when they came to France, or protesting with the peasants and the Zapatistas in Mexico – effectively, our demands are the same. Of course there are different points of view in Via Campesina – it's the exchange of opinions and experiences that makes it such a fantastic network for training and debate. It's a real farmers' International, a living example of a new relationship between North and South.

Shouldn't the anti-globalization movement oppose globalized forms of military power – NATO, for example, as well as the WTO?

That's more complicated. It's not to say that one shouldn't fight against NATO. But behind the military conflict there is often a far more cunning and destructive form of economic colonization going on, through the programmes imposed by the IMF and World Bank – opening regions up to the multinationals, dismantling public services, privatizing utilities. In Sarajevo in the mid-nineties, for instance, there were people in the French military contingent who weren't officers at all but representatives of the multinational, Vivendi – originally Eaux de France. They spent their whole time studying the water mains and the infrastructure. When the fighting was over, they were on the spot to offer their services in reconstructing Bosnia's utilities. Today, it's Vivendi that runs Sarajevo's water system, as a private service. It's a form of economic domination that we're seeing throughout Latin America, Africa, Asia and elsewhere.

We do need to denounce the role of the sole military superpower as world policeman. But its economic dominance is more important. There tend to be anti-war protests against particular conflicts, rather than around militarism as such. There was quite a big mobilization in France against the Gulf War, although it wasn't easy since it was a Socialist government that was prosecuting the War. But the way the West struck simply in order to control the oil was so brazen that it did generate real protest. In Bosnia and Kosovo, the situation was much more ambiguous. There was a lot of debate inside the movement between those who opposed the NATO intervention and others who said, quite rightly, that Milošević's regime was a rotten, red-brown affair – the old Stalinism in Serb national dress. And people had known what was going on in Kosovo for years. There was a lot of discussion as to what form resistance and solidarity should take. But for me, there can never be a good war. As soon as you reach that stage, it is inevitably the people who lose. I was against both forms of military intervention, just as I oppose the American bombardment of Afghanistan.

What is your attitude to the anti-globalization 'republicanism' of Chevènement, which has had its reflections in Left thinking elsewhere: Benn in Britain, for example?

I had a public debate with Chevènement on French radio when I was at the anti-globalization conference at Porto Alegre last January. It came down to an opposition of two completely different points of view. Chevènement thinks that the borders of the nation-state can serve as a rampart against globalization. I believe that's an illusion. Multinational corporations, multi-lateral accords on investment, free-trade rules operate on quite another level, over and above national frontiers. To say one can have a strong state makes no sense in this context. It just gives people the mirage of a satisfactory form of protection. As Interior Minister, Chevènement was responsible for implementing the most restrictive immigration policies, abrogating the basic human right to freedom of movement. Closing the

frontiers does nothing to resolve the fundamental issue at stake in immigration – the inequality between North and South.

*Surely the one state whose power hasn't lessened in the face of these multilateral
accords is the USA?*

Of course the US completely dominates the IMF and the World Bank, and
its will is hegemonic within the Security Council. But the US government,
in turn, is just a tool of the big companies. Its political function is simply to
relay the economic interests of the major firms – which is why, in the last
elections, many people didn't see any choice between Bush and Gore. Ralph
Nader's campaign highlighted the real nature of American politics. Candidates are effectively elected to be the representatives of financial or industrial
groups. The system is entirely at the service of economic interests, which
retain the real power. One can see this happening in detailed ways at the
level of the federal administration: the power of the multinationals imposes
itself directly on the running of the machine. The US state functions as a
motor of support for them, institutionally and ideologically. But neoliberalism is not just an American preserve. It goes right across the board – Europe
or America, governments of the Right or Social Democrats. In their
negotiations with the WTO, there has been no difference between the
current EU commissioner for trade – Pascal Lamy, a member of the French
Socialist Party – and his predecessor Leon Brittain, a British Conservative.
The same thinking – *la pensée unique* – really is hegemonic everywhere today.
It's not just *la pensée américaine*. We need to pay attention to its proponents
within our own countries, rather than see only the Stars and Stripes.

Jospin came to power promising a more radical agenda than either Blair or Schroeder — what's the balance sheet?

There is scarcely any difference between the economic programmes of the Right and Left – if one can call the Socialist Party that. For example, there's been no attempt at a genuine reduction of the working week, just a series of negotiations within each sector. They're trying to take a middle path. They could have gone much further. Now, with their eyes on next year's elections, the PS have been trying to recover votes on the Left by making a show of interest in the autonomous movements. But it's just at the level of talk. They're doing nothing about the movements' programmes at the level of policy. At the WTO talks in Doha the French government will be right behind the EU positions. The main question in the legislative and presidential elections next spring will be the percentage of abstentions. A lot of people have been very disappointed in the policies of the Union of the Left – and they don't necessarily recognize themselves in the hard-left candidates, who will get a few votes in the first round. Chirac and Jospin offer no real choice between alternatives. Their vision of society is the same. We're moving increasingly towards a situation where economic logic is stronger than any political will. Party leaders simply adjust to the prevailing wind. The Confédération Paysanne is not calling for a vote for any of the parties. I myself wonder whether one should vote at all.

There has been talk of your standing in the presidential election yourself?

Never. That's not my role. In fact, it's a condition of membership of the Confédération Paysanne that you cannot stand in an election. Curiously enough, the first person who said I was thinking of standing in the presidentials was Daniel Cohn-Bendit, just after Seattle. A few days later the Socialist Party repeated it – as if the aim was to break the social movement by saying: they do all this just to serve as a trampoline towards a political party, or to enter office. As if one couldn't have an autonomous movement with a logic of its own, acting as an oppositional force outside the established political domain. I would never see it as my role to act like the leader of a political party, as a professional representative who takes responsibility out of other people's hands. The aim of a social movement or a union like ours is to enable people to act for themselves. The economy has become an autonomous sphere today, imposing laws of its own. If we are going to create a new politics we have to understand this.

You went to the Israeli-occupied territories this summer, to demonstrate with the Palestinian farmers. What did you learn about the situation there?

First of all, I experienced the reality of the Israeli military occupation of Palestine – that it really is a war of colonization. They're trying to impose an apartheid system on both the occupied territories and the Arab population in the rest of Israel. They are also putting in place – with the support of the World Bank – a series of neoliberal measures intended to integrate the Middle East into globalized production circuits, through the exploitation of cheap Palestinian labour. Along the frontier with the occupied territories, they're setting up the same sort of enterprise system you see along the Mexican–US border. So there is a very acute economic dimension to the conflict. The UN resolutions need to be implemented. But there also needs to be a radical reorientation at the economic level, that would offer a viable future to the Palestinians.

The financial press has been triumphantly announcing that September 11 has put paid to the anti-globalization movement. What is your assessment – did the terrorist attacks in the US 'change everything'?

Underneath, nothing has changed. The world situation remains the same. The institutions are unchanged. And the anti-globalization movements, too, are still here. With the bombardment of Afghanistan, we are seeing the domestic propaganda needs of the United States being elevated to war aims, inflicting revenge on an innocent people already suffering miseries of deprivation, while threatening further destabilization in that part of the world. There is also no doubt that the US wants access to oil wells outside the control of OPEC, and may have its eye on reserves in the ex-Soviet republics of Central Asia. The position of the Confédération Paysanne has been: 'No to Taliban, No to Terrorism, No to War'.

We also see a new awareness, born of the economic crisis, of the need for regulation and public intervention. In that sense, the logic of globalization is more on the defensive now. The critique of neoliberalism that we have been developing over the last years is more valid than ever after September 11. But the response of most of the states who've signed up for what they call the 'war against terrorism' is to call for an expansion of neoliberal policies, as if that could resolve the inequalities between different countries, or social layers. They have understood nothing. September 11 should have been a chance to take stock of the sort of social and ideological costs this regime has been exacting, and to call for its radical reform. Instead, they are seeking to reinforce their global domination, escalating the dangers of wider international conflict. As neoliberalism increases the balance of misery in the world, it just augments the numbers of those desperate enough to throw themselves into fanatical, suicidal attacks against it.

APPENDICES

APPENDIX 1

FROM SOCIAL PROTEST TO A CRITIQUE OF SOCIETY: SOME KEY DATES IN THE HISTORY OF THE FARMERS' CONFEDERATION

The history of the Farmers' Confederation is also the history of the small and average farmer. Guided by the values of equality, it has rejected the effects of intensive farming, with its production-centred excesses. It was born from the protest movement and challenged the hegemony of the FNSEA, until then the sole agricultural union, which had been founded on the myth of farmers' unity and on the structures established by the Vichy government.

THE 1960s AND 1970s: PROTESTS AGAINST THE SOCIAL EFFECTS OF THE MODERNIZATION OF FARMING

21 July 1969: The regional Federation of Western Farmers' Unions, chaired by Bernard Lambert, publicly challenges Sicco Mansholt, the European

Farming Commissioner, and denounces the plight of small farmers forced into modernization.

April 1970: Bernard Lambert publishes '*Farmers and the Class Struggle*' and together with young protesters from the CNJA, sets up the 'Worker-Farmers' movement, which is strongly influenced by the ideology of May '68; their journal, *West Wind*, is first published in November 1970, and at two successive CNJA congresses (1970 and 1971) they come close to winning a majority.

May 1972: The 'Worker-Farmers' organization breaks away from the CNJA just at the moment a long strike in the milk delivery business starts, followed by a strike of milk producers in three Departments of Brittany, demanding 'a living wage from the price of milk'.

August 1973: The first large gathering in the Larzac in support of the farmers fighting against the extension of the military camp takes place. The presence of the striking Lip workers (who had taken over the production and sale of the watches they manufactured) marks the beginning of a unity between farmers and workers which continues throughout the 1970s.

Spring–summer 1974: Many farmers' demonstrations are organized against the overproduction of beef and pork. These are often led by 'Interpaysanne', a pressure group within the FNSEA.

Summer 1974: A second gathering in support of the 103 Larzac farmers is held.

1974–75: Significant property disputes take place in western France; empty farms are occupied and given over to young farmers; workers on strike and farmers occupying their farms lend support to each other, and are met with fierce repression.

1974–80: Farmers join in anti-nuclear protests against new power stations in Brittany.

February 1976: A crisis in the wine-growing industry. A demonstration in Montredon (South of France) against massive imports of Algerian and Italian wines leaves two dead (one farmer and one policeman).

Summer 1976: Drought breaks records in the North and West of France, plunging farmers who had been 'modernizing' their farms into deeper debt. Two years later, in September 1978, a hunger strike by Jean Cadiot against the Crédit Agricole Bank leads to the formation of a protest group in defence of 'farmers in debt'.

February 1978: Two large farmers' demonstrations are held. The first, organized by 'Interpaysanne', mobilizes 7,000 people in opposition to a 'joint responsibility' tax on the price of milk. The second, organized by the Worker-Farmers, attracts 2,500 people to the trial of a large veal feed firm which had cheated a large number of its farmer customers.

March 1978: The FNSEA expels oppositionists who were active in 'Interpaysanne'.

THE 1980s: FROM A CRITIQUE OF INTENSIVE FARMING TO AN ALTERNATIVE, DEFINED AS SUSTAINABLE FARMING

September 1980: A boycott of hormone-fed veal is organized after veal farmers revealed that they were forced by their co-operative to use banned hormone-based foodstuffs. Bernard Lambert and the Worker-Farmers lead this campaign and attack intensive methods.

June 1981: The National Confederation of Farmers' Unions (CNSTP) is set up by groups and local unions who have broken with FNSEA policies and grouped around the Worker-Farmers. Their first Congress, focused on the issue of intensive farming, takes place in Paris.

April 1982: The FNSP (National Federation of Farmers' Unions) is set up.

February 1983: Elections for the Chambers of Agriculture confirm the existence of the CNSTP and FNSP as autonomous movements outside the FNSEA; between them they poll some 15 per cent of the vote, and have a presence in over seventy Departments.

April 1984: The EEC imposes milk quotas, which are strongly opposed by the CNSTP and FNSP. Both organizations are torn between protecting small farmers doomed by this policy from the outset and demanding an alternative quantum policy.

April 1987: The Farmers' Confederation is created from the fusion of the CNSTP and FNSP, and an organization called Rural Hope 76. Its motto is: 'For small farming and the defence of its workers'.

November 1989: The Farmers' Confederation organizes a conference in Paris on the 'Estates General of the rural world', attended by more than fifty rural mayors.

February 1990: Together with defence groups for farmers facing financial difficulties, the Farmers' Confederation organizes a conference on 'Problems in Farming', at which a new organization, Farming Solidarity, is set up to co-ordinate local groups.

THE 1990s: A NEW AGRICULTURAL POLICY KNOWN AS SUSTAINABLE FARMING

February 1990: A decree on Farming Unions Representation is published, giving the Farmers' Confederation the right to sit on a limited number of commissions on farming policy at local and national level.

1991: A successful mobilization against the construction of a giant battery chicken farm in east Paris.

1991–94: Between the 1992 introduction of the CAP (the EU's Common Agricultural Policy) and the 1994 GATT agreement drawn up in Marrakesh, the Farmers' Confederation continues to denounce the free market, but is unsure what tactics to adopt.

April 1993: Following protests by small-scale farmers, a Charter (see Appendix 2) is drawn up; it becomes public in December 1998.

February 1995: Elections to the Chambers of Agriculture confirm the increase of Farmers' Confederation influence: the Confederation obtains 26 per cent of the vote (the FNSEA's share falls to below 60 per cent).

February 1996: The EU reinforces its embargo on the use of growth hormones in cattle breeding.

June 1997: The Socialist government, back in power, passes a law giving farming unions full rights of representation on farming policy bodies. The Farmers' Confederation participates in drafting a new law on farming, and is consulted on all major policies by the Ministry of Agriculture.

January 1998: A symbolic act of destruction of genetically modified seeds takes place in Nérac (Lot-et-Garonne); three union representatives (José Bové, René Riesel and François Roux) receive a suspended sentence of several months' imprisonment. Campaigning against genetically modified food starts.

August 1999: In Millau, farmers, including José Bové, are imprisoned for the symbolic action of dismantling a McDonald's. This marks the start of an international mobilization against the WTO, culminating in the breakdown of negotiations in Seattle in November 1999.

APPENDIX 2

THE CHARTER FOR SUSTAINABLE FARMING: FARMING THAT RESPECTS THE FARMER AND MEETS THE NEEDS OF SOCIETY

FARMING TO SERVE SOCIETY

Such farming fulfils a number of needs:
* *for food*: consumers increasingly want food to taste good and be wholesome; they want to know how it is processed;
* *to renew activity in rural areas*: until the 1950s, farmers made up 50 per cent of the rural population, with farming at the centre of rural life;
* *to consider* the environmental and land-management aspects in the goods and services provided by farming;
* *for respect towards the natural habitat*: to maintain quality and diversity in farming.

To respond to these needs as a whole, farming produces two types of goods: commercial goods (foodstuff) and non-commercial goods (environment, landscape).

Sustainable farming offers the degree of quality demanded by consumers, and involves *three distinct aspects*:

- a social dimension based on employment: sustainable farming allows for a larger number of farmers to be employed;
- *economic efficiency* to create added value, crucial to maintain maximum numbers in employment;
- *respect for consumers and nature*: preservation of ecological balance, bio-diversity, landscape and food quality.

These various dimensions depend on farmers' individual choices, and on the overall political framework – agricultural policies can either encourage or hold back the effects of sustainable farming.

PROCEDURES AND AREAS

Sustainable farming is defined by a 'procedure' and an 'area'.

The 'procedure' involves the direction of farming, as expressed by the 'ten principles of sustainable farming' (see below).

The 'area' is the specific land allocated to farming activities.

Without the following two conditions, farming cannot develop:

- A positive political framework which supports traditional farming instead of encouraging intensification.
- Farmers' individual choices.

The drawing up of a Charter is important; its functions are:

- to conduct a complete analysis of the type of farm involved, where progress needs to be made;
- to act as a support framework for training, organization and development;
- to draw up suggestions for agricultural policies;
- to put the issue of sustainable farming at centre stage.

THE TEN PRINCIPLES OF SUSTAINABLE FARMING

Principle 1: Distribution of production to allow for the maximum number of people working as farmers – the right to produce includes the right to work and the right to an income.

Principle 2: Solidarity with farmers in Europe and the rest of the world.

Principle 3: Respect for nature. Nature must be preserved to ensure the continuity of its use by future generations.

Principle 4: Making the most of abundant resources, and protecting those that are rare.

Principle 5: Transparency in all relations of purchasing, production, processing and sale of agricultural produce.

Principle 6: Ensuring the good quality, taste and safety of produce.

Principle 7: Maximum autonomy in the running of farms.

Principle 8: Partnership with others living in the countryside.

Principle 9: Maintaining the diversity of the animals bred and plants grown. For both historic and economic reasons we must preserve the biodiversity of the land.

Principle 10: Always bearing in mind the long-term and global context.

These ten principles, taken together, are the basis of sustainable farming; while each is a prerequisite, only taken together do they represent the totality of the approach.

APPENDIX 3

USEFUL ADDRESSES

Confédération paysanne
81 avenue de la République
93170 Bagnolet
tel. 01 43 62 04 04
fax 01 43 62 80 03
e-mail: confpays@globenet.org
Website: http/www. confederationpaysanne.fr

Coordination paysanne européenne (CPE)
rue de la Sablonnière 18
B 1000 Brussels
tel. 00 32 2 217 31 12
fax 00 32 2 218 45 29
e-mail: cpe@agoranet.be

Via campesina
Rafael Alegria M.,
Secretario operativo internacional.
Apartado postal 3628, Tegucigalpa
MDC, Honduras
tel.& fax 00 504 220 12 18
e-mail: viacam@gbm.hn

Coordination pour le contrôle citoyen de l'OMC (CCC-OMC)

44, rue Montcalm

75018 Paris

tel. 01 46 06 46 30

fax 01 46 06 41 07

Website: http//www.citoyens.net

Association pour la taxation des transactions financières pour l'aide aux citoyens (ATTAC)

9 *bis*, rue de Valence

75005 Paris

tel. 01 43 36 30 54

fax 01 43 36 26 26

e-mail: attac@attac.org

Website: http//attac.org